D0723836

TECHNICAL KNOWLEDGE IN AMERICAN CULTURE

History of American Science and Technology Series

General Editor, LESTER D. STEPHENS

Technical Knowledge in American Culture

Science, Technology, and Medicine Since the Early 1800s

Edited by
HAMILTON CRAVENS,
ALAN I MARCUS, AND
DAVID M. KATZMAN

The University of Alabama Press

Tuscaloosa and London

Library of Congress Cataloging-in-Publication Data

Technical knowledge in American culture : science, technology, and
medicine since the early 1800s / edited by Hamilton Cravens, Alan I
Marcus, and David M. Katzman.
    p.   cm. — (History of American science and technology)
Includes bibliographical references and index.
ISBN 0-8173-0793-1 (alk. paper)
1. Science—United States—History. 2. Science—United States—
History—19th century. 3. Science—United States—History—20th
century. 4. Medicine—Unites States—History. 5. Medicine—United
States—History—19th century. 6. Medicine—United States—
History—20th century. 7. Technology—United States—History.
8. Technology—United States—History—19th century. 9. Technology—
United States—History—20th century. I. Cravens, Hamilton.
II. Marcus, Alan I, 1949- . III. Katzman, David M. IV. Series.
Q127.U6T37   1996
306.4'6'097309034—dc20   95-19084
                                          CIP

British Library Cataloguing-in-Publication Data available

We dedicate this book to our students:
those we have taught and those we seek to reach

Contents

Acknowledgments

"The American Career of Jane Marcet's *Conversations on Chemistry*, 1806–1853," originally appeared in *Isis* 82 (1991): 9–23; and "Diagnosing Unnatural Motherhood: Nineteenth Century Physicians and 'Puerperal Insanity,'" "Race-ism and the City: The Young Du Bois and the Role of Place in Social Theory, 1893–1901," and "Introduction: American Studies and American Science: An Analysis" originally appeared in *American Studies* Fall 1989 (special issue), the last in a very different form. We thank *American Studies* and its sponsoring organization, the Mid-America American Studies Association.

TECHNICAL KNOWLEDGE IN AMERICAN CULTURE

HAMILTON CRAVENS AND ALAN I MARCUS

Introduction
Technical Knowledge in American Culture: An Analysis

Science, medicine, technology. These activities conjure up various images for students of American culture. In our own time they seem related to society and culture as problem and solution, as threats to the social fabric and remedies for it. Technical knowledge appears crucial to modern life, whether it is medical, scientific, or technological in character—or whether it represents any other kind of technical discourse. After all, it is the publicly shared discourse of the professional elites that appears to run society, economy, polity, and culture. Is such knowledge a part of the larger culture, or not? In the case of the law or military strategy and tactics, the answer appears commonsensically clear: such expert knowledge is a part of American society and culture. But what about technical knowledge—science, technology, and medicine, that is?

In presenting the essays that comprise this book, we argue for two interrelated propositions. The first is that modern technical knowledge emerged in western (and American) history in the early nineteenth century, thus sundering public discourse and the public domain; what has taken its place has been a plethora of privatized discourses and arcane bodies of knowledge. We believe that this is a common thread that unifies the essays that follow: what was more general knowledge about the natural world and people's attempts to understand or use it became transformed in the post-Napoleonic world into a congeries of technical discourses or bodies of knowledge about plants, animals, the weather, the composition of substances, the planet, and other natural forces, as well as the causes and cures of disease, and the like. Our second proposition flows naturally from our first: science, medicine, and technology have emerged as distinct bodies of knowledge no less than have

I

law, theology, and business administration—or any other fields, for that matter. And it is chiefly as intellectual, not social, constructs that we wish to consider science, technology, and medicine. For reasons that will rapidly become clear, we wish to consider them within the rubric of technical knowledge.

In the last thirty years the history of science, technology, and medicine have emerged as seemingly disparate fields of history because their leading champions have assumed a social history model; that is, they have argued that it is the *social,* not the intellectual, history of science, technology, and medicine that has mattered. Put even more simply, the "social impact" school has insisted that what makes the history of science, technology, and medicine important is its impact on society. Apparently the social impact school has swept all opposition before it. It is uncommon now to find a work on science, technology, or medicine that pays much attention at all to the intellectual and ideological activities of scientists, doctors, or technologists, save as these activities confirm or modify particular hypotheses about the social impact of the scientist, the engineer, or the health care professional. In a word, the social impact school is firmly entrenched in scholarly circles today and needs no defense, least of all from us.

What we propose in this volume is to turn this social impact model on its head. We wish to look at science, technology, and medicine not as social activities, and especially not in terms of whatever sociological categories might appear useful at the moment. Rather we insist that what scientists, engineers, and health care professionals have done and do intellectually and ideologically is important to study. It is, in short, the *knowledge* that they use and create that marks them off as distinctive practitioners and professionals and that makes their history remarkable and worthy of historical investigation. So long as science, technology, and medicine remain defined as social activities and defined in sociological terms (e.g., social forces, social communities, social impacts, and the like) it is difficult to group them together for the purposes of investigation, description, and interpretation. As social (and sociological) activities and processes they all appear, or have been made to appear, to have radically different histories. What is the common thread among them? This is indeed the problem that haunts many advocates of social history in America today: can we move toward synthesis, toward a more general understanding of the American past, or more precisely, of the several distinct pasts that constitute America's history?

Technical knowledge, for our purposes here, may be defined as that special and arcane knowledge that scientists, engineers, doctors, and their colleagues and coworkers use in their work. Furthermore, in these essays we assert that such technical knowledge, as with any other kind of discourse or knowledge, has its ultimate origins in the larger culture itself. In short, the interpretive unity that we seek in history today is not to be found in detailed social matrices but in cultural constructs, which in turn are mental constructs. Such mental constructs are the basis of the social matrix of a given time and place no less than the underlying notions about what Michel Foucault dubbed the "order of things," those tacit agreements among contemporaries about how the world works in a given age. We do not wish to revive the consensus school of American history of the 1950s or the old-fashioned *Geistgeschichte* school of the later nineteenth and early twentieth centuries, in which the historian identified a particular idea and then attempted to find examples of it in various lines of activity. This is too crude a technique. We insist rather that contemporaries agree on a common interpretation of how the world works in a given age, so that they can understand one another despite the often widely divergent interests—social, economic, political, ideological, and the like—that they hold and represent. We are interested in the history of cultural forms and intellectual structures *wherever* they might be located in a given age and culture, without regard, that is, to the (presumed) social matrix. By the same token we are not all interested in trendy (or even passé) versions of "structuralism" as grasped, no matter how foggily, on either side of the Atlantic.

It is an odd but incontestable fact that science, medicine, and technology in American culture, as objects of study, have been at best of marginal concern to students of American culture. There has been some interest in the problem of technical knowledge in American culture, to be sure, but almost always among historians, not other students of American Studies and culture. Thus in the interdisciplinary field of American Studies, none of the field's major journals, such as *American Quarterly, American Studies,* and *Journal of American Studies,* have done more than to carry occasional articles on these subjects. Usually the authors of these single studies were budding specialists in the history of science, medicine, or technology, not in American Studies or in another branch of American culture studies, such as American intellectual history or American literature. These authors would then return to their home turf, whatever that might have been, where they developed their careers. Occasionally scholars bridged the gap between the study of science, technology, and medicine in America and American culture with books, including such distinguished scholars as Merle Curti and Stow Persons. Such work did not generate a school of American cultural studies scholars studying the history and development of technical knowledge—of science, medicine, and technology, that is—in American culture. Nor has there been much intellectual cross-fertilization between historians of science, medicine, and technology on the one hand and American culture scholars on the other.

Why have these two great wings of American culture studies—science and literature—not taken much interest in one another?[1]

Models for study and cooperation do exist. Two schools of thought emerged by the mid-1960s. In his *The Structure of Scientific Revolutions* (1962) Thomas S. Kuhn, a philosopher and historian of science, outlined a program for cultural historical studies that revived both Neo-Kantian and Hegelian philosophy in more up-to-date costume, and, in turn, amplified the post–World War II idealist program that Alexander Koyre and others in philosophy and history of science championed. Kuhn argued that the scientific community's investigatory and interpretive efforts were directly influenced by shared, often tacit, models of nature, or, in Kuhn's argot, paradigms. Through various means of socialization—commonly education and apprenticeship—the established practitioners incorporated the younger generation of scientists into the community. In such ways was the communal paradigm passed on to succeeding generations. Most scientists spent their careers working out the paradigm's puzzles, taking the discrete facts with which they worked and fitting them into the larger paradigm at the appropriate points.

Because science in past and present are never the same, Kuhn had to account for changes in science's history. He explained the dynamics of change from one "paradigm" to another as internal—thus change stemmed from actions within the community of investigators. Specifically there came a point at which the discovery of new facts as anomalies with the old shared paradigm created such a burden of conflict and stress that a revolution took place in the perceptions of the research community's investigators, and, in short order, the community created a new paradigm in which the new and old facts could be comfortably juxtaposed and reconciled with one another. Thus there was "progress" through scientific revolutions.

What Kuhn offered was a program in which the ideas of science remained crucial and important within a social community whose sociological (as distinct from psychological and cognitive) dynamics were never spelled out; his focus remained on scientists' ideas. And for those interested in the social history of ideas, or the sociology of knowledge, his work appeared fresh and exciting. He stressed the importance of the scientific community as social medium in which the ideas and actions of scientists took place and as the mechanism (or congeries of mechanisms) through which the "progress" of science took place. This seemed ample grist for the mills of historians, social scientists, and other specialists in American cultural history and studies. Moreover, Kuhn appeared to make the history of science more accessible to nonscientists, specifically those trained in the humanities and social sciences, by insisting that the internal dynamics of a social entity—the scientific community—was the locus of science's history.[2]

The other school of thought was even more accessible to students and scholars of American culture studies than that which Kuhn represented. It might be thought of as the moralistic or even metaphysical school of the history of technology, a congenial enough perspective for many humanists and social scientists. Its most articulate champion was, if anything, better known to American Studies scholars than Kuhn—Leo Marx. His *The Machine in the Garden: Technology and the Pastoral Ideal in America* (1964) offered a work that was far more amenable to the interests and abilities of American Studies scholars and students than was science and its history. Marx's book rapidly became a classic in American Studies, one of several such other seminal works in this field, beginning with Henry Nash Smith's *Virgin Land: The American West as Symbol and Myth* (1950) but also including R. W. B. Lewis's *The American Adam* (1955), John William Ward's *Andrew Jackson: Symbol for an Age* (1955), and Marvin Meyers's *The Jacksonian Persuasion* (1957)—all texts of the "myth and symbol" school of American Studies. And, again, work in that interdisciplinary field was symptomatic of that in all areas of American cultural studies.

Those who worked in this vein used literary and historical materials (with emphasis on the literary) to explain various aspects of the essence of the American experience or of the national character, which was the result of that experience. Thus these scholars took an essentialist approach to American culture: the essence of America, past and present, was this or that main theme or issue, such as the economic or frontier interpretations of American history as offered up by Charles A. Beard and Frederick Jackson Turner. For those of the myth and symbol school, the tension between nature and civilization spelled the essence of the American experience. That tension, of course, in turn derived from modernist literary criticism and philosophy—thus its metaphysical affiliations. As for the myth and symbol school's approaches and interpretations, the school's guru, Henry Nash Smith, explained that myth and symbol stood for larger or smaller units of the "same kind of thing, namely an intellectual construction that fused concept and emotion into an image"; thus myths and symbols were collective representations, "not the work of a single mind."[3]

Marx stressed what he dubbed the pastoral ideal in America, and its concomitant dichotomy of nature versus civilization, as the essence or central meaning of America, past and present—a Hegelian dialectic if there ever were one. The pastoral ideal, he argued, had been used from the age of discovery to the present century to define the meaning of America. Europeans were dazzled by the possibilities of a fresh, new, virgin world that the New World offered—including those of withdrawal from the troubled and problematic Old World.

Marx interpreted the pastoral ideal as the central cultural construct and symbol of American culture past and present as it changed from William Shakespeare's *The Tempest* to F. Scott Fitzgerald's *The Great Gatsby*. Its chief manifestation was what he called the machine in the garden, a wonderfully effective metaphor for the metaphysical tension between nature and civilization. Thus the machine stood for industrialization, and the garden represented bucolic or pastoral America. The machine in the garden signified the invasion of technology and industrialism into pastoral America. "When the Republic was founded, nine of ten Americans were husbandmen," Marx argued, whereas "today not one in ten lives on a farm. Ours is an intricately organized, urban, industrial, nuclear-armed society." Marx devoted considerable space to the years 1800 to 1860 when, he insisted, many Americans were torn over the machine's intrusion into the garden, and when many seemed to believe that, despite this potential contradiction, all would be well, for the upshot would be what Marx labeled a "middle landscape," in which the pastoral fused with industrialism in new and presumably beneficial ways, what one of Marx's students had termed "workshops in the wilderness." Hence the early American linking of good land and honest labor could now include virtuous manufacturing and commerce as well. Yet Marx insisted that this soon changed. After 1860 the "middle landscape" became increasingly unrealistic. The dream of "a rural nation exhibiting a happy balance of art and nature" became chimerical, a mere rhetorical bromide rather than a social blueprint, "an increasingly transparent and jejune expression of the national preference for having it both ways," thus enabling the nation to continue to define its purpose "as the pursuit of rural happiness while devoting itself to productivity, wealth, and power."[4]

If Marx had carried an old message, the nature-civilization tension, into our own time, he had nevertheless dressed it up in appropriately contemporary language and with language and notions that were trendy for their time—the mid-1960s. In particular he had outlined a stinging moralistic indictment of technology; and in that rested whatever freshness his ideas may be said to possess. At least since the 1920s scholars, writers, and commentators in the United States had viewed technology—and, by extension, medicine and science—as a positive social force that influenced society and culture profoundly. This was part of perceiving technology—and science and medicine too—as solutions to social problems. What Marx injected into the professional discourse of American culture studies was the post-1950s notion, drawn from contemporary culture and society, that technology was a metaphysical entity and, therefore, morally good or bad. Of course Marx, like an increasing number of his fellow Americans, especially those on the left, saw technology as bad, as a social problem even more than a social solution. In this way did Marx advocate the labeling of technology as autonomous force from a moralistic as well as a metaphysical or essentialist perspective.[5]

Initially Kuhn and Marx attracted much attention, and in certain ways they still do. Yet perhaps ironically, relatively little scholarly or monographic work has flowed or derived from theirs, so that if these are schools of interpretation, they seem oddly lacking in graduates and disciples, at least in the direct sense. The difficulty would appear to rest not so much in the ability of Kuhn and Marx to attract attention and, at least, to be quoted and cited in professional discourse and chitchat. Rather the problem arises with regard to published works of scholarship whose authors derived their work from Kuhn or Marx, for such a literature is, for all intents and purposes, virtually nonexistent. How can two authors whose work has been so widely discussed have been, nevertheless, by this yardstick, not seminal at all? This seems baffling in the extreme and requires explanation. Perhaps there have been flaws in the execution of their respective programs.

For many in the history of science and cultural history in the 1960s, Kuhn's insistence that ideas could be linked with communities seemed to suggest a resolution to the internal versus external—or ideas versus social experience—debates that then raged in these academic circles. As a general formula, Kuhn's thesis provoked considerable discussion worldwide, and he has received mountains of mail, requests to speak, and all manner of other evidence of his currency of the last quarter-century. It has been, in many ways, a phenomenal and almost unbelievable academic success story. But Kuhn's arguments, as wondrously helpful, even liberating, as they were, remained only a topic of discussion and controversy, not a program of action. To paraphrase Kuhn, he did not create a paradigm that others could use as the basis for further practice.

The problems, alas and alack, were multiple and nettlesome. What was a paradigm? Was it a microstatement, a statement organizing material at some middle level of consciousness (or a variety thereof), or was it a model of the world? Were there important paradigms that were not statements about nature? And there seemed to be a pre-paradigmatic phase in the history of science before there were universal paradigms. Could it really be that there was something called pre-science and then, after a certain point, something that was science? Clearly this was a positivistic statement. Presumably many scientists could easily accept it. But was it useful for a historian or social scientist? How could historians and social scientists properly create objective criteria for distinguishing between "pre-science" and "science," let alone shape the notion of a paradigm into an effective and productive research agenda? And, for that matter, how could such a perspective be used in other areas of cultural history studies? Thus how could it be applied to politics and political theory, literature and drama, social thought and public policy, or other nonscientific activities? Were there pre-paradigmatic periods of literature and politics, for example? Or was this just another version of pre-sentist whiggery, modernization theory, or other simple-minded linear explanations? And what did this say about Kuhn's notions of science?

Nor was this all. Kuhn gave great authority to the notion, not original with him or his work, that there were "internal" and "external" factors in the development of science, meaning quite literally internal and external in relation to the scientific community itself. This also encouraged thinking of them as dichotomous, clearly not helpful if one wished to mediate these angles of vision as had seemed possible with Kuhn's work initially. Despite some overtures to the contrary, as when he discussed how paradigm shifts and revolutions occurred, at bottom Kuhn thought of science as progressive, additive, and cumulative. Humans might be fallible, but the knowledge of science somehow was not, or at least was less so. He did not take up the platonic argument that knowledge existed independently of human history, but on closer examination he might as well have insofar as his potential influence among those trained in the humanities and social sciences were concerned. Hence Kuhn's ultimate effect among American Studies scholars was not to encourage the study of medicine, science, and technology, at least if by encourage one means direct inspiration and discipleship. It was true that some American culture scholars, such as the late Gene Wise, did pick up on the notion of cultural paradigms, but its harvest in American culture studies for any kind of scholarly investigations seemed slim pickings indeed. And that was the net yield in the history of science as well. This is not to say that Kuhn had no generalized influence among historians of science or among a wide variety of scholars interested in cultural studies. He did and does. But, again to paraphrase Kuhn, his work was nonparadigmatic: it could not be used as the basis of further practice.[6]

The legacy of Marx's work has been problematic in a different way. Among American Studies and American literature scholars, for example, his approaches to historical and literary sources have remained popular, especially for undergraduate teaching. It is perhaps revealing, however, that scholars in literature and American Studies eschew the particular myth and symbol approach method for their investigations. In 1972, Bruce Kuklick published what became a famous methodological critique of the myth and symbol school in the pages of *American Quarterly,* the premier journal of American Studies. In it he accused its members—including Marx, of course—of a variety of serious errors that compromised their scholarship and interpretations. These errors included reading back into ancient sources the concerns of modern intellectuals through a confused Cartesian dualistic approach to the history of ideas and, furthermore, assuming that one could read from popular culture to the motives and ideas of members of an entire society. These were, for Kuklick, disastrous gaffes. They eliminated whatever intellectual benefit the myth and symbol school might have been able to offer to cultural studies.[7]

It is notable that Kuklick did not criticize the larger implications of Marx's work concerning the moral dimensions of technology and American culture. Nor, for that matter, did most other students of American culture, regardless of their disciplinary orientation. Yet from a historian's standpoint, or more precisely from that of the pure historicist, Marx's work constituted a metaphysical and moralistic critique of the baneful role of technology in American culture that closed off scholarly discussion. And in that particular sense Marx's legacy differed sharply from Kuhn's, whose contribution seemed to provoke discussion without end. There was another difference. Kuhn was the ultimate scientist-insider on several different levels of meaning. Marx, on the other hand, was no technologist-insider. He viewed technology from the outside, as a self-professed "humanist" and not too subtle moralist and metaphysician. According to Marx, technology was an intrusive force, monolithic, all-powerful and bad. Once it intruded into the garden that he claimed that America was, it was all over. To put Marx's argument negatively, as it were, technology was not a multifaceted human activity in which human beings were involved and in which they participated.

Thus it was simply impossible to historicize Marx's argument. One could write what appeared to be a historical narrative, but it would not, because it could not, recognize the otherness of the past, or, more pointedly, the otherness of the many pasts that have constituted the past. By insisting that the minute industrialism came, Americans could never have their rural past again and that the unfolding future was both ineluctable and bad, Marx helped articulate a particular research agenda for the investigation of the relations between technology and society. This research program would stress the depiction of the impact of technology on society. And it would do so in obviously presentist and activist ways. As Marx would have it, the intrusion of the machine into the garden was the beginning of our own time. This was not a useful perspective for someone interested in the past for its own sake. Eventually several writers noticed that, in the words of Howard P. Segal, there was more than one middle landscape, so that there were more possibilities in America's past than Marx had ever conceded.[8]

Marx did not inspire derivative scholarly studies of technology and culture among the multiple constituencies of American culture studies for various reasons. One appears overriding: Marx's book simply closed off discussion. There was no way, outside of positing more middle landscapes, that a scholar could follow up with more investigations if that person worked from a cultural history perspective and with the kind of training commonly available for such work. The only avenue led to new middle landscapes. But these seemed relatively limited in number and severely circumscribed in explanatory power. They were but brief stopping points on the inevitable journey to the awful present. In the hands of Marx and his devotees, they were static, ahistorical categories—caricatures of the past, really—and could not serve as a useful guide to scholarly investigation and interpretation.

By the middle to late 1970s the legacies of Kuhn and Marx seemed clear. In each case the imprint was a generalized one that helped set the tone for further conversation and speculation. A rising generation of historians interested in the history of science, medicine, and technology in American culture was coming to the fore. In books and articles these scholars drew attention among historians of American science and medicine to such topics and subjects as the development of the scientific community in the nineteenth century; to the work of the medical and public health communities of the late nineteenth and early twentieth centuries; to the influence in America of the ideas of "great men" such as Lamarck, Pasteur, Koch, Darwin, and Freud; to the importance of the social processes of professionalization and certification; to the social relations of such fields as physics; to the development and institutionalization of Darwinian theory, the evolution of medical education, the successes of campaigns to stamp out impure foods and drugs and to fight terrifying diseases; and to the involvement of the federal government in science. Here the generalized influence of Kuhn was strong, as in the emphases on the communities and the ideas of science and medicine. As for Marx's influence on historians of American technology, there were, not surprisingly, virtually no studies based on his specific approach, as had been the case with Kuhn and historians of science and medicine. But—as historians of American technology published books and articles on technological institutions, the engineering profession, space flight and aeronautics, the so-called American system of manufactures, the role of warfare, industry, and social life—the humanist perspective that Marx had championed, in which technology could be judged morally and metaphysically, often dominated the agenda of discussion and research, sometimes openly, as with David F. Noble's *America by Design* (1977) or less obviously as with Merritt Roe Smith's *Harpers Ferry Armory and the New Technology: The Challenge of Change* (1978).[9]

Here we must recognize the popularity of the new social history, which has reordered research and teaching agendas in studies of American culture, historical and otherwise. What began in the late 1950s and early 1960s as a call for a more quantitative (and, therefore, representative) approach to historical research both in research methods and in conceptualizations of the relative importance or relevance of historical phenomena in the past—most commonly in political and economic history—became transformed by the late 1960s into the new social history, a new and interesting program for discussion of contemporary and scholarly issues. The new social history had political as well as scholarly agendas and goals. This did not differentiate these historians from earlier generations of historians so much as emphasize a changing of the guard within the profession's ranks.[10]

The new social history of science and the new social history of America's groups—and thus the new American cultural studies scholarship, manifestly including that of American Studies—constituted parallel schools of thought. And in turn they shared their origins in a chorus of concerns that were current in the 1950s: the belief in ineluctable social forces as well as the notion that the whole of American civilization had a distinct moral essence that could be judged. There appeared in the 1950s a language of protest against conformity, belonging, and togetherness, as in the hip and beat movements, that mushroomed in the 1960s and 1970s into a full-blown discourse of protest against oppression in the past and present, as in the popular moral movements for civil rights, feminism, and against the Viet-Nam War. In the new, post-1950s age, increasingly Americans from many walks of life acted and spoke as if there was an infinity of dimensions, proportions, relations or, more simply, as if life was individuated, with no larger whole, only an endless number of parts.

Ours is an age in which we think of society as a collection of individuals—individual persons, single issues, an infinity of perspectives—and we cannot, try as we wish, regain that holistic sense of society that existed between the 1920s and the 1950s. In some circles, this is known as postmodernism, or the perception that the world itself is divided into an infinite number of dimensions, proportions, and shapes. That kind of thinking about the order of things in the world has been symbolized by such fascinating cultural artifacts as William H. Whyte, Jr.'s famous study of organization men in the 1950s, with its witty, savage attacks on conformity, and also David Riesman's *The Lonely Crowd: A Study of the Changing American Character* (1950), and even that quintessential 1950s motion picture of conformity, *The Man in the Grey Flannel Suit,* adapted from Sloan Wilson's 1955 novel. More recently such thinking has been reflected in forthright (and often highly moving) narratives of white racist oppression of nonwhites in the 1960s; in the 1970s in a variety of calls for less is more, such as that of E. F. Schumacher; and in the 1980s in what Tom Wolfe called the "plutography"—among other representations.[11]

For historians, this agenda has devolved into the social history of science, technology, and medicine, which has meant, in the main, institutional, social, and political history. Historians of American science, however, have also been sensitive to the issues of race, gender, and social oppression, so that there have been works on women and members of minority groups as their lives have been touched by science, medicine, and technology, ranging from discrimination against women scientists to the eugenics movement or other manifestations of scientific racism. For scholarly interpreters of American culture, past and present, the categories of race, gender, and class have been crucial, as for the historians, but unlike among historians, among American studies scholars there seems little interest in the history of scientific institutions or the politics of science. The connecting thread is there, to be sure, between the two groups of scholars. It is precisely the assumption that the social matrix (or social matrices) order and control all human behavior and thought—all of society and culture.[12]

Given the assumption that the social matrix determines thought and action, then presumably investigations into the social history of science—and of medicine and technology as well—would not appear to be relevant or important to American culture studies scholars who were looking into the workings of a different part of the social matrix (or a different social matrix) pertaining to gender or minority group status and history. This has indeed been the fate of American intellectual history, which used to be a broad field in which one attempted to understand important issues in the culture as a whole. There still are some scholars who use what they consider the tools and materials of intellectual history to address large issues in American history, such as Gordon Wood on the American Revolution, or Linda Kerber on the Republican mother, or Carl Degler on Darwinism in American culture; but this probably does not represent intellectual history to most historians as it might have in the pre-1960s era. And from that mainstream point of view, the field has become merely the social history of the intellectuals. And more is the pity.[13]

Ultimately for many scholars the standard or mainstream social history interpretation assumes different loci for different groups in the social matrix of the past because of their differing "importance," so that in our own time, assumptions of historical "relevance" and "importance" are bound up with assumptions and presumptions concerning location, meaning the width of the gap, if any, between distinct historical actors or groups thereof. And in our own time, in many quarters in history and American culture, scholars and activists alike assume a plebiscitary or "democratic" mentality about historical "importance"—a consequence, if there ever were one, of our individualistic way of understanding the world. From this perspective, it is not surprising that many scholars and activists regard such activities as science, medicine, and technology as marginal to their interests in the new social history, for these activities would appear to be distant from what seems important and relevant, and, in any event, are dominated by elites and involve substantial and arcane intellectual life—hardly the stuff of the new social history and its research and policy agendas. If most in American cultural history who were committed to the new social history were to have any interest in science, medicine, and technology at all, presumably it would be in the *social* history of science, medicine, and technology. Given the public discourse of our own age concerning the relations between society and science, technology, and medicine, in which it is assumed that such technical expertise victimizes ordinary citizens, it is hardly surprising that the social history of science, medicine, and technology, for most scholars, is or should be the study of their impact upon society. Ultimately this is why Marx's *Machine in the Garden* has resonated with so many scholars and activists: it is and always has been a very timely metaphysical or moralistic condemnation of the machine as despoiler and oppressor of humankind. Clearly this is a very popular attitude among many Americans, whether they are scholars or activists or not. Such an attitude would make Kuhn's approach to the social history of ideas not especially exciting to many.

In truth, ours is an age in which expertise of any kind, but certainly technical knowledge, is often challenged and criticized more than it is revered or trusted. And what a convenient position for one to take who believes that it is the social impact of technology or science or medicine that matters most. After all, from the social history perspective, what the scientists, engineers, and doctors were as part of the social fabric appeared trivial—because they were elites, and elites were anathema, according to social history doctrine—as compared with the awesome havoc they could wreak upon the social order. And it would not be so intellectually daunting for a scholar trained in the humanities and social sciences to take up such discussions from a social history *cum* moralistic point of view. The technical or professional discourse of science, medicine, or technology would be, for most, too forbidding. From this perspective, it would not make much sense to regard science, medicine, or technology as windows onto the past, no more or less "important" than any other kind of coherent human activity in the past, even though that perspective might suggest how a historian or other specialist in American cultural studies might penetrate the mysteries of scientific, medical, and technological activities in American (or any) culture and society in the past.

Nor does that exhaust the matter. To invoke William James, there have been varieties of social history approaches in the history of American science, medicine, and technology. Even within the field of the new social history there have been enormously fruitful possibilities, provided one is willing to forego such assumptions as linear development and social-eco-nomic determinism to undergird one's larger explanatory frameworks. In addition to the quantitative strains within the new social history there have been qualitative ones as well. Often such studies have been of elites and their various constituencies, and the knowledge—broadly defined, of course—that has constituted the larger discourses that these social groups have used and reacted to may be thought of as parallel cultural constructs to those in science, technology, and medicine. Thus we have had, within the new social history, a mixture of traditions, including the quantitative and the qualitative, and often we have had most interesting and promising stud-ies for the ultimate and yet to be accomplished reconstruction and unifica-tion of American cultural historical studies, notably studies of streetcar sub-urbs, of urban ghettoes for blacks and immigrants, of attempts to understand the feelings and attitudes of those on the bottom of the socioeconomic-po-litical hierarchy, and of folklore, especially among those groups marked by gender, race, and powerlessness. And not all new social historians teach that the relations of elites and masses have been the attempt by the former to impose social control or, more trendily, hegemony over the latter. There are some new social historians who stoutly claim that they have moved beyond the study of victimization. There are possibilities for a unified history of American culture, and, more specifically, a rich, complex, and sophisticated history of knowledge, including technical knowledge, in American culture.

Similarly, there are promising leads and suggestions within the history of science, medicine, and technology as well. Here we refer specifically to the genre of historical explanation in which one writes, in fairly straightforward fashion, about the history of scientific or medical or technological institu-tions or professionals in these fields. When this has been connected to the ideas of the *dramatis personae,* and even to the larger culture, the results have often been gratifying. And the parallels of this to the qualitative tradition within the new social history are plain as day.

Yet this is not a lead that has been followed much, especially in recent years. More common has been discussion of the social impact of science, technology, or medicine—in other words, how these phenomena impact upon society, as with Stephen J. Gould's *Mismeasure of Man* (1981), in which he insists that the inegalitarian ideas of scientists have led to the horrors of racism and sexism. Another version has been to discuss social context as formative: what a scientist or doctor or technician thinks is but a reflection of who he or she is in society.

The difficulty with such social history agendas—especially the latter two—is the problematic assumptions undergirding them. Thus it is simply not self-evidently true that science, medicine, or technology can be thought of as autonomous social forces beyond our control, after the fashion of Leo Marx, or as ultimately true platonic statements, as Thomas Kuhn would have it. Neither approach offers a useful, constructive guide to the study of the past on its own terms. Nor, for that matter, is it a sustainable assumption that one's social background or individual psyche, or both, constitute a predictable guide to one's attitudes and behavior. This might be the case here and there; but often such remains a problematic and slippery statement that closes off painstaking research and hard thinking with what amounts to the invocation of thunder and lightning as magic![14]

Thus some social history approaches to cultural studies cannot provide what their champions so ardently claim for them. Like some cultural history perspectives, they cannot provide the grand interpretive synthesis that many scholars are calling for with increasing frequency. In a finely differentiated social matrix such as the new social historians often posit, there can be no concept of unity save the social matrix itself, which in the hands of many has been reified into a metaphysical unity, such as "industrialism" or "modernization" or the like. Thus Marx has closed off discussion before it can begin, and Kuhn has provided no practical guidance for participants to end discussion. As for other approaches, those that contain problematic assumptions and deny the otherness and pastness of the past offer little guidance to the reconstruction of the human past. Nor is this all. Often such advocates violate a simple truth. People are not ants. People carry on a cultural life of ideas, in the broadest sense, that is not reducible to social matrices and assumptions about social location and space. This is no mere claim of woolly-headed philosophers and humanists, but the ultimate discovery of the human sciences in our own century. We ought to pay it some decent respect and set aside the confused and bewildered reductionist schemes of late nineteenth-century psychobiology and economic theory, which are unfortunately an inspiration for much contemporary thinking about history and human affairs—unfortunately, we say, because so much of that thinking is so muddled, ahistorical, and downright silly.

There are genuine possibilities that bode well for the sophisticated and useful reconstruction of the problem of technical knowledge in American culture or in any culture. Here we can glimpse another way of approaching the phenomena of science, medicine, and technology that might prove helpful for all students of American culture. As suggested above, this tradition exists in the literature of the history of technical knowledge in the United States. Those who have deployed this perspective have stood the standard social history formulation on its head. They have asked what the influence of American culture has been on science, medicine, and technology. They have used as their point of departure, in other words, those common, widespread notions in the culture that provide a particular age with its unifying sense of the order of things, or taxonomy of natural and social reality, and how these intellectual constructs have influenced such complex and multi-faceted phenomena as science, medicine, and technology in the multiple pasts that add up to the sum of American history. They have embraced a profoundly historicist approach to the reconstruction of the past, which the mid-nineteenth-century Swiss scholar, Jakob Burckhardt, pioneered in his classic study *The Civilization of the Renaissance* (1859). What this approach has meant is a commitment to understand the otherness of the past, its differ-entness from the present, which has consisted not so much in events as such, on the surface, but in those deeper structures of perception and explanation, those interpretations and definitions that contemporaries in a particular age make of the world around them and that do constitute finely differentiated mental and cultural historical phenomena providing an age its unity and coherence to contemporaries. In any age, contemporaries have to agree on "the order of things," even though they widely disagree on many important issues and interests. That is another fundamental "truth" that the human sciences have yielded in the twentieth century.

The number of distinguished examples from the extant literature of the history of science, medicine, and technology in American culture is not enormous. In his classic *The Pursuit of Science in Revolutionary America* (1956), Brooke Hindle covered a wide canvas of scientific activity in eighteenth-century America and discussed men of science of widely differing back-grounds and interests. But he also demonstrated the intellectual unity of the age, the essentially mercantile character of the scientific enterprise, a notion entirely congruent with those organizing the larger culture and society, and how that manifested itself in the scientific institutions of the age. He also convincingly narrated the establishment of the first homegrown scientific institutions in the western hemisphere during and after the Revolutionary War. Daniel J. Boorstin, in *The Lost World of Thomas Jefferson* (1948), and George H. Daniels, in *American Science in the Age of Jackson* (1967), covered well from this perspective their topics for the early to middle nineteenth century. Unfortunately the most noted works for the mid nineteenth cen-tury have not followed this promising lead, but for the late nineteenth and early twentieth centuries there are other books in this genre, including Henry D. Shapiro's *Appalachia on Our Mind* (1978), which may be considered a landmark in both history and American culture studies. Shapiro located the causes and support for successive scientific ideas about Appalachia and its denizens in contemporary cultural notions about America and Americans. Thus ideas of scientists—here social scientists—were the products of cultural notions.[15]

Such an approach offers certain advantages to the scholar of American culture. It underlines the point, which in today's discourses unfortunately bears continuous reiteration, that what people do in a specific situation is the product of how they think of the realities of nature and civilization, rather than of impersonal, suprahuman social forces. There is neither need nor reason to introduce such reified abstractions as social forces. This distorts the realities of the human condition. This has the effect of undermining dispassionate scholarship by "picking sides." And a deterministic model of the past such as this tends to undermine an activist agenda (although activism is not near and dear to our hearts).

Yet there is more. The approach that we urge here makes the study of science, medicine, and technology in American culture accessible to scholars trained in the humanities and the social sciences. To be sure, the humanities or social sciences scholar must learn the technical discourse of the discipline or field of knowledge to be studied. Upon occasion this could include the humbling prospect of acquiring in midlife a competence in, say, calculus or thermodynamics. Perhaps one would be fortunate enough to have a clever adolescent in the household who could offer pointers. But ultimately this problem would recede, as would all others. And the professional expectations and rewards are fulfilled. The principle of scholarship remains the same for all fields of knowledge: one must understand the activity under scrutiny. Those who have studied literature, whether poetry or prose, or art or drama or cultural communities will find that the study of, for example, Darwinian concepts or of particular notions of disease is inherently no more difficult than is any other subject that they have mastered. One does have to learn, and to understand fully, the technical concepts, those building blocks that people in any human activity use to construct it. But in principle it is no more difficult to write about notions of fluid dynamics than of poetry; at bottom all human actions and constructs are grounded in ideas, and ideas that are ultimately accessible. One may have to learn a technical discourse; fair enough. Usually the greater challenge, however, is of understanding the otherness of the past, of how differently people in a given age understood their own times from the ways we would grasp them. And this approach enables the scholar to avoid the dangers of presentism, of placing human action and belief in the wrong historical period. This approach will facilitate historicization; the past will not be merely a reserve of amorphous historical facts, some of which are relevant to current concerns and others not. It will have structure, differentiation, and order. Above all, it will be not one past but many. In this way will truly distinguished scholarship and effective activism become genuine possibilities.

The essays in this volume do exemplify the method—the impact of culture upon knowledge—outlined above that appears so promising and illuminating. Thus the first several essays clearly identify dominant cultural notions of the early to middle nineteenth century and the rise of democratic culture, and relate these phenomena to science, technology and medicine. Later essays dramatically demonstrate the all-sufficiency of those larger cultural notions in the creation of technical knowledge in the age of hierarchy and "system" from the 1870s to the 1920s, as with the racial social science of W. E. B. Du Bois or the racial nutrition schemes of Ellen Richards and W. O. Atwater and their allies, or the invention of technical knowledge about the mentally retarded, about airports and urban "metroplexes," or even about animal growth hormones, such as diethylstilbestrol (DES). Our authors insist that technical knowledge, in one way or another, is always shaped and informed by the larger culture and, within limits, by the society that is produced by that larger culture. Here then is the unifying thread of cultural studies, whether of technical knowledge or of anything else: it is not the social matrix (or matrices) of the new social historians. Rather it is the common, usually tacit understandings of the order of things that contemporaries in any age share so as to be contemporaries of one another. Such consensus, if that be consensus, on the order of things did not preclude conflict over specific interests in culture, society, polity, and economy. Ours is, after all, a thinking species that creates not merely societies (many of the so-called higher animal species do that), but cultures of a world of ideas, symbols, languages, logic, classifications of the world, and the like. We ignore such realities in dealing with history at our conceptual and evidentiary peril.

Thus when we read each essay we come to understand that in a given age, different groups of Americans confront problems and in doing so redefine the world around them. This volume's authors have located, or fixed, each essay in a period of time, surrounded and circumscribed by a series of notions common to that time. These notions were peculiar to Americans at that time and constrained what then contemporaries saw and did. It was a distinctly American invention or construction of knowledge, and this "knowledge" informed and gave meaning to science, technology, and medicine. Indeed, an era's science, technology, and medicine reflect that era's cultural notions. That recognition provides tremendous benefit to all students of American cultural history and studies. It frees the study of technical knowledge—here, the study of science, technology and medicine—from the monumental tombs in which the followers of Kuhn and Marx regally consigned and finally buried them. It places these subjects directly within American culture, accessible to citizens and scholars as in fact they should be, and as part of that unique, changing construct. Rather than progenitors of culture, the kinds of technical knowledge—science, technology, and medicine—become keys and guideposts to the revelation and explanation of American culture. Ultimately they emerge as vital clues for understanding the American past.

PART ONE

The Rise of Democratic Culture, 1800–1870

Introduction to Chapter 1

The first several articles in this volume clearly identify dominant cultural notions of the early to middle nineteenth century—the rise of democratic culture, that is—and relate them to technical knowledge—to science, technology, and medicine. In her absorbing essay on the important Ohio Mechanic's Institute, Judith Spraul-Schmidt argues that the institute's founders thought it an instrument of public policy, an institutional solution to the most pressing problem of that age: incivility. As most historians of the early nineteenth century have noted, although it was officially an age of widespread optimistic belief in the perfectibility of humanity and the progress of civilization, many Americans nevertheless feared that their nation, as the land of opportunity, still had too many denizens who remained illiterate and uneducated, which was clearly, in Spraul-Schmidt's words, "unfortunate for the affected individuals but disastrous for a democratic republic." Thus the problem of incivility—of too widespread illiteracy and lack of education, that is—subverted the concept and the very functioning of a country built on republican virtues. Character development, after all, was an important factor in success; and success itself was the product of individual initiative. Without literacy and education, or civility, the wondrous promises of the democratic republic could never be fulfilled.

Cincinnatians founded the Ohio Mechanic's Institute in the fall of 1828 and won for it a state charter the next winter, thus underlining its role as an institution for the entire state, and not merely for the Queen City of the West, as its citizens fondly referred to it. As its main spokesman, John D. Craig, and other founders believed, the institute would diffuse useful knowledge to all citizens, thus promoting civility. At that time, none of Cincinnati's rival cities in the west had such an institution, which simply made Cincinnatians even more proud of their city. In the 1830s the institute's officers embarked on an ambitious campaign to diffuse knowledge of all kinds to the citizenry at large through an expanding library, a curriculum of regular courses, and other aids for gaining useful knowledge in the arts and sciences, including mathematics, natural philosophy, chemistry, and natural history. The institute also offered courses in French and German language, and even studies in English grammar. By the early 1830s, Spraul-Schmidt insists, the institute was succeeding in its program of enabling young men to become so well educated and civilized that they could join the ranks of the institute's directors—perhaps the supreme accolade of being civilized.

In the later 1830s, the Ohio Institute's directors embarked on a new program, that of sponsoring annual fairs in which all the material objects manufactured or made within the city were shown, a wondrous way of promoting Cincinnati's yeomen and its manufactures. Yet no sooner had the fairs begun than the institute itself began to redefine itself in ways that represented a drastic shift in purposes and constituencies. No longer fascinated with individual self-improvement, suddenly the Ohio Mechanic's Institute altered its program to represent the needs of particular groups within the citizenry. The annual fairs advertised categories of items manufactured by Cincinnatians. The institute's curriculum became strikingly altered within a few years to instruct carpenters, plumbers, and other operative mechanics in technical knowledge, not that general knowledge useful to all society. By the late 1840s the shift was complete, as Spraul-Schmidt's arresting and subtle narrative shows us.

Spraul-Schmidt thus insists that the Ohio Mechanic's Institute was but one example of the many self-culture institutions Americans founded in that era as the means of providing opportunities for Americans to make themselves into civilized individuals who would be free to pursue commerce and trade without undue interference from an oppressive or corrupt central government. Certainly the movements for public schools, subscription libraries, lyceums, penitentiaries, asylums, houses of correction, and even eleemosynary institutions more generally, all reflected the same underlying deeper cultural notions about the organization of natural and social reality—that once all individuals in society had been civilized, no further reforms or alterations in public policy would be needed. The republic would be secure.

The institute thus illuminates the history of technical knowledge and the larger culture in several ways. First, as a manifestation of its era, it represented the underlying notions of a free society populated by civilized individuals who were free to pursue commerce and trade untrammeled by the force of tradition, monopoly, or other forms of irrational rule. Second, as an institution for self-help, it was a conservator and purveyor of technical knowledge in a variety of formats. And third, as Spraul-Schmidt suggests, its ultimate fate suggests much about how America shifted from one era to another.

JUDITH SPRAUL-SCHMIDT

The Ohio Mechanic's Institute: The Challenge of Incivility in the Democratic Republic

The Ohio Mechanic's Institute was formed in late 1828 as "an Institution in the city of Cincinnati, for advancing the best interests of the Mechanics, Manufacturers and Artizans, by the more general diffusion of useful knowledge in those important classes of the community." Offering public lectures and regular courses of instruction and possessing a circulating library and reading room, the institute took its place alongside the new common schools, academies, lyceums, and subscription libraries. Each of these vibrant, burgeoning institutions was designed to increase opportunities for learning. Each functioned as "a sort of universal self-and-community improvement association," providing its participants with intellectual and character-building enlightenment. In an age in which character development—civility— seemed an important factor in success and in which success seemed the product of individual initiative, providing opportunity for individuals to cultivate these traits proved appropriate public policy. It would enhance the nation's citizens and their prosperity.[1]

This thread was embedded within the national fiber. The American republic's future depended on its yeomen—its citizenry—achieving civilization. It led to the creation of self-culture institutions to give a broad segment of society the means for developing the individual's inherent potential. But all was not rosy. While the flurry of reform activity in the 1820s and 1830s, as historians have noted, reflected the optimistic spirit of the age—the belief in the perfectibility of humanity and the progress of civilization—the effort's breadth implies something dire. An annoying and widespread conviction had surfaced that America as the land of individual opportunity fell short of expectations. The fervor with which Americans conducted their

24

campaign to foster civilization addressed that point. The problem of incivility loomed large and threatened the new nation. Contemporaries maintained that too many Americans remained illiterate and unenlightened, unfortunate for the affected individuals but disastrous for a democratic republic. It promised to undermine the concept of a country built on republican virtues. Yet the incivility of the 1820s and 1830s did not seem without remedy. Establishment of institutions to diffuse knowledge more generally appeared effective counteragents.

Perhaps the common school movement best demonstrated the idea that Americans needed to take positive, even public action to make the rudiments of education available to a wide spectrum of future citizens. Americans experimented in this period with new instructional methods and opened academies of learning to improve the education of the young during their formative years. Cities throughout the nation acted as innovators by urging state legislatures to provide support for schools open to the public free of charge, seeking legislative backing for the statewide establishment of free schools. But the problem of incivility did not confine itself to children, as contemporaries saw it, nor did the possibilities for redress of the deficiencies. Subscription libraries, discussion societies, mechanic's institutes, and the lyceum proliferated to take up the challenge.[2]

Indeed, the conviction that the state of civilization in America could be raised through individual exertion was extended even to those exhibiting overtly uncivil behavior. For example, creation of the new American penitentiary system suggested that even the criminals were redeemable. At the Auburn prison in New York and the Eastern State Penitentiary of Pennsylvania, "in solitude [the prisoner] was expected to read the Bible, to meditate upon his condition, and to develop mentally and spiritually, until he might, upon obtaining his freedom, lead a new and different life." Perpetrators of crime, if given proper direction, could assume the responsibilities of citizenship. Juvenile offenders did not belong in prisons, but in their own "houses of correction" or "reformation" where they learned the proper way to behave in a civilized nation and acquired some specific skills, such as carpentry or weaving. Voluntary associations opened orphan asylums and houses of employment for the female poor to teach manners, morals, and trades, while offering a salubrious atmosphere to prepare their clients for readmission to the larger society. Asylums for the blind and deaf served similar functions. Between 1817 and 1836 state governments and private groups established asylums calculated to supply the insane "with appropriate medical and moral treatment" in Philadelphia, New York, Connecticut, Massachusetts, Kentucky, Virginia, and South Carolina.[3]

Nowhere was the desire and need to civilize individuals more powerful than in cities. These densely populated places made incivility more apparent. But it remained their contemporary definition as sites in which individuals gathered for the opportunity of pursuing commerce and manufacturing that tipped the balance. City government existed to provide that opportunity. And in the parlance of the day, success was dependent upon civility. It devolved to cities, therefore, to offer citizens those civilizing opportunities. Accordingly, much of the early nineteenth-century self-culture campaign was urban-based, and Cincinnati as one of the nation's fastest growing and principal western cities proved no exception.[4] It emulated the older, more established eastern seaboard cities, confident that Cincinnati would dominate the West. In the words of a local booster, what had been in 1815 "the late humble village of Cincinnati" had already by 1826 "advanced to the rank and opulence of a city," and concerted efforts of its citizens seemed entirely capable of raising it further. Not surprisingly, Cincinnatians looked favorably upon reports that a new kind of institution, the "mechanic's institute," in the past four years had taken hold in New York, Philadelphia, Boston, and Baltimore and "proved beneficial to the interests of the community, both in respect to wealth, morality, and happiness."[5]

The desirability of an institution that could promote those ends and the apparent reasonableness of doing so through a program of "diffusing useful knowledge" encouraged Cincinnatians to take steps to establish their own mechanic's institute. They saw the opportunity for adding to their list of urban cultural institutions a sign that the city actively promoted its citizens' opportunity and therefore well-being and that it confronted incivility, defined as the inadequate diffusion of the light of learning through the population. This too stemmed from a lack of opportunity. Contemporaries argued that individuals would naturally pursue their own success. That some failed to achieve it or even seek it implied that they lacked the opportunity. Institutions to provide the requisite knowledge were not reaching some, and new, different institutions were required. Accordingly, "mechanics and other citizens of Cincinnati" gathered on October 28, 1828, to organize "a Mechanic's Institute." The assembly listened to reports of similar institutions in the "Atlantic cities," learned that none had been formed in Cincinnati's rival cities—St. Louis, Pittsburgh, Louisville, or Lexington—and resolved to plan such an institute.[6]

John D. Craig, a nationally known lecturer in "natural philosophy" and future head of the United States Patent Office, explained to Cincinnatians the objectives of a mechanic's institute. That sort of institution demanded consideration from "any community desirous of encouraging the useful arts and manufactures and [at] the same time, of guarding against the moral turpitude and degradation of human character" that follows from "mental darkness." He asserted that many Americans were unenlightened, a serious problem for the country. "The state of the arts" of belles lettres, scientific learning, and production of goods "forms the principal criterion, by which one estimates the degree of civilization among any people," Craig argued. If the nation's level of civilization was deficient, its citizens must create institutions to spread opportunity to develop that state. Without such a campaign, even most profound events would "in a great measure, prove a dead letter, to the great masses of people." Teaching civilization was never a simple task. It must be, Craig contended, "brought home to their doors, and down to their comprehension, by a living voice, and ocular demonstration." Only in that fashion could the formerly unenlightened grasp the "laws and economy of nature" and apply these principles in their own lives as citizens in an orderly civil society.[7]

Craig also held that a mechanic's institute could benefit those already enlightened, especially "all youths of genius and exemplary conduct." In a refrain reflecting the era's emphasis on perfectionism, he maintained that these young men could develop themselves further, acquire knowledge and mastery "to improve and extend their respective arts beyond any known limits: and raise those who practice them to that rank in society to which their utility entitles them." Each of these youths would become, like Benjamin Franklin, a "philosopher mechanic." Together they would constitute a Jeffersonian-like "aristocracy of talent," recognized community leaders. New practical inventions and commercial prosperity would come from these young men of native genius, another reason why each state government needed to establish a mechanic's institute. "Here as in many other improvements," Craig concluded, "the people must take the lead and their rulers will follow them." It required citizens to form institutes and then seek state aid. Each institute was in the public interest because it would, Craig noted, "shed its influence—its Illuminating Influence, over the community at large, by disseminating rational entertainments and mental embellishments: . . . [and] induce the rising generation to suspend the movements of folly and dissipation, to acquire knowledge and manly accomplishments."[8]

Craig's sentiments struck a responsive chord, and Cincinnatians drew up and approved a constitution at that same meeting. It laid out the manner in which the Ohio Mechanic's Institute expected to "promote the mechanic arts, to offer general enlightenment and to promote the best interests of the community." Membership was available to any Cincinnatian who would agree to its provisions and pay a nominal membership fee, suggesting that virtually the entire citizenry fit within the term "mechanic" in the broad sense of the institute's name and purpose. So too did the institute's emphasis on public lectures, courses, and library study. Institute leadership was vested in a member-elected board of directors. Any member could serve on the board. The institute also reached out further and granted "minors, sons, and apprentices of members, if of good moral character, . . . the privileges of attending any lectures of the Institute, and . . . the use of the library, agreable [sic] to the regulations thereof, on payment of fifty cents per annum, the parent or guardian being responsible for the same."[9]

The act of forming such an institution called attention to Cincinnati's claim for primacy in the state of Ohio. The granting of a charter of incorporation by the Ohio Legislature in February 1829 demonstrated that the institute remained a state institution. Its name, the Ohio Mechanic's Institute, reflected state authority. The legislature formed the institute as a "body corporate and politic . . . competent to contract and be contracted with, to sue and be sued . . . with full power to acquire, hold, possess, use, occupy and enjoy, by devise or otherwise, and the same to sell, convey and dispose of, all such real personal estate as shall be necessary and convenient for said Institution." The singular possessive form of the term "Mechanic's" indicated that the Institute was for and of the individual mechanic, defined by the board of directors and others to include virtually anyone—entrepreneurs, factory or small business owners, craftsmen, journeymen and laborers—engaged in the prosecution of the mechanic arts.[10]

Securing an education in the early nineteenth century was the responsibility of each individual and did not imply collegiate instruction. Everyone could use free time to read and gain knowledge, Cincinnatians were reminded, and an 1828 newspaper communiqué went so far as to calculate the number of hours per week—12—and per year—677—that even the busiest man should have been able to manage. "If a mechanic then spends the time we have allotted him in literary and scientific pursuits," the calculation continued, "at the end of twenty years [or 1,128 days and four hours of reading] he will have devoted more time to literature and science than any of our educated men will have in six years" of formal schooling. The "mechanic" who consciously educated himself would "obtain a proper influence in society"; he would as a mechanic "join the educated part of society [which] gives laws to the rest." Such a man would recognize that "cultivating his mind" and "acquiring a skillful knowledge of the human character are as necessary to him as a knowledge of his tools," and he would gain "his due proportion of influence with the world."[11]

Its philosophical underpinning and governmental authority outlined in detail, the institute lost no time in beginning its classes and public lectures. "Dr. Locke's Lectures in Geometry" were among the earliest courses. John Locke studied medicine and botany with Benjamin Silliman at Yale and had only recently come to Cincinnati. A member of the prestigious American

Philosophical Society, Locke became an important ornament and exemplar of the Ohio Mechanic's Institute. Like other institute classes of that first year including arithmetic and chemistry, Locke's geometry class met in the evenings during the work week and revealed the natural world's general principles.[12]

Regular class participants could also attend public lectures, which the institute offered on Saturday evenings. Always advertised in the local press, the lectures were open to "the citizens generally and the ladies particularly," usually without admission charge. The lectures provided information on a wide variety of topics and a forum for generating interest in institute membership.[13] From the first, the public lecture figured prominently as the institute's best means of securing "the greatest knowledge, to the greatest numbers, at the least expense." Lecturers would volunteer their services and share their expertise with the wider assembly. Speakers addressed a wide range of topics because "the improvement of the mind, other circumstances being the same, is in proportion to the number, variety, and extent of objects with which it is conversant." Many lectures focused on branches of "natural philosophy" explicating the physical world, especially heat and chemistry. "Mechanical philosophy," or explanation of the workings of machinery, also figured prominently,[14] as did speeches on the laws of civilization. "Optimism," education, religion, and the Constitution of the United States proved fit topics. "A discourse on the methods, means, and necessity of educating the BLIND," "the remote, predisposing and exciting causes to intemperance," and "the importance and practicability of a more general diffusion of legal knowledge, among our young men" also appeared on the lecture program. Each was part of the general campaign to extend useful knowledge through the population, while the diversity of topics suggests just how broadly contemporaries conceived of that category.[15]

Offerings of a single participant, Dr. Daniel Drake, epitomized the scope of knowledge deemed appropriate to an industrious individual. Drake called on his experience as a physician to lecture on anatomy and physiology, respiration and the circulation of the blood, and the physiology of the nervous system. He drew upon his membership in the American Philosophical Society and long years as correspondent on the West's natural history to speak on climate and atmospheric pressure. Even "phrenology" and "physical and intellectual education appropriate to the ladies" proved fit subjects for Drake's inquiry. As important, Drake's presence at the podium of the institute symbolized that each citizen could "ris[e] in the world—and obtain . . . an influence and an honest name by means of his own exertions." Rising in the world seemed open to anyone, even "the ladies." The institute established specific classes and formal courses of instruction exclusively for females, but unlike the members' evening classes, they were held in the afternoon.[16]

The institute's directors sought to tap into the publicity generated by these lectures to increase membership, and they aimed specifically at "practical artisans." Almost from the start, they worried that the membership fee might prohibit some from joining. Because the directors were "sensible of the advantages which the community may derive from the success of the Institution, and fully impressed with the belief that its prosperity depends on the encouragement afforded to it by the mechanics and artists of our city," the board permitted mechanics to deposit articles of their workmanship in lieu of membership fees. The institute would display these articles as examples of local production and once a year offer them for public sale. The balance in excess of membership fees would be turned over to the maker.[17]

In its second year of operation, the institute acquired its own building, the failed Bazaar built by Frances Trollope in the city's core, on Walnut between Third and Fourth streets. It soon included a reading room where members gathered for congenial discussion and perused current periodicals. Members were impressed that the reading room "was furnished with many valuable scientific and miscellaneous works, besides a good collection of Periodicals." Donations of a "club of gentlemen" provided the room with newspaper subscriptions and "about forty of the most valuable American and foreign periodicals." The Philosophical Society of Cincinnati also provided a valuable assist. That club shared with the institute its library as well as "instruments of investigation," especially of the physical sciences, "including Mathematics, Natural Philosophy, Chymistry, and Natural History."[18] The institute took complete control of this facility in 1832 when the society disbanded.[19] The earlier organization's books and journals provided a solid foundation for the institute's library. Open in the afternoons and evenings to serve laboring members, the institute library made circulating materials accessible to members, while exacting fines for overdue books to insure their timely return.[20]

In 1831 the institute reaffirmed its commitment to young workers by promising to afford "to those young men whose avocations allow them but limited periods of leisure, the means of employing that leisure in obtaining such knowledge as may be most useful to them." It was to these young men that the institute offered its regular courses of instruction. These varied yearly from arithmetic, mathematics, chemistry, and geometry to natural philosophy. The institute also provided language courses in French and German and studies in English grammar, and even some older members interested in developing their full potential sat in. These diverse offerings led a commentator to suggest that "the Institute holds forth to young men of this city, greater advantages than perhaps any other" Cincinnati organization. More than one hundred young men had taken institute courses by 1831, a fact that caused the board of directors no end of pride. It announced proudly that two of the young men who had diligently availed themselves of all the benefits of the institute, from the period of its commencement, had "at the recent elections, been included in the number of the directors, and the other members of the board anticipate with pleasure the period, when their own places may be filled by others who will have obtained the most important portions of their education in the Institute, and who will thereby be the most competent to direct its operations." Already, its directors argued, the institute showed itself an appropriate means of fostering natural leadership and expanding the number of civilized, productive individuals in America.[21]

The Cincinnati community received an opportunity to show its support of the institute's youth programs when in 1833 a twenty-five-cent admission charge was levied on a public lecture to defray equipment costs. Locke's lecture, "Electro-Magnetism," drew newspaper backing as the media printed alongside the institute's notices their own endorsement that "a more laudable object of expenditure could not be proposed." One paper even published a letter from a correspondent who insisted that civic-mindedness demanded that the city support "scientific exhibitions [for] young men who, regardless of the paltry and frothy amusements of theatres, coffee houses, &c., strive with all their gettings to get understanding."[22]

Locke's lecture appeared a viable fund-raising tool for student equipment to an audience already familiar with the institute and its functions. And the institute needed the money because its program of lectures, classes, and library offerings stretched it to the limit. By the mid-1830s it reported itself "like all new beginnings in the West, labouring under pecuniary embarrassments" and unable to meet the payments on its building. For several years, the institute rented space and by the end of the decade began to charge admission for lectures. Its decision in 1838 to hold a public exhibition, or fair, was coupled with a "large and splendid BALL," to fund that and other institute affairs. "Mechanics and Artists of the city and vicinity" were called upon to help plan the gala, and the press urged all citizens to purchase the five-dollar tickets for the February 26, 1838 "Grand Mechanics and Citizens Ball." Actual attendance at the ball, one editor explained, was less important than demonstrating "the liberal feelings of the entire community" by supporting the institute financially. The fundraiser proved hugely successful, added $2,812.24 to the institute's treasury, and enabled it to proceed with plans for its first annual fair.[23]

The fair seemed yet another means of providing opportunity to Cincinnati's yeomanry. Exhibitions of local products had been successful both in Europe and in "large eastern cities of our own country," proponents explained, exciting emulation "among ingenious mechanics and artisans, and . . . making known to the community at large their respective merits." One reason for fast-growing Cincinnati's successes, institute supporters implied, was its attractiveness to these yeoman. The institute's program of courses, lectures, and library offerings was but a single example. Its activities provided chances to create better handiworks and to engage in the advancement of civilization. The Ohio Mechanic's Institute did not stand alone. As early as 1829, Cincinnatians had formed a Chamber of Commerce to advance the city's reputation and prosperity. That institution diffused knowledge, gathering and disseminating "all the intelligence essential to just views of business" for "mutual advantage" of buyers and sellers. More than mere market prices and quantities were included the chamber's store of information as its proponents created dossiers on "the character and standing of mercantile men," a move necessary because "Cincinnati may now be considered a great commercial emporium." To maintain and advance that rank, institutions like the Chamber of Commerce and the Mechanic's Institute attempted to insure commercial honesty as well as quality workmanship. By 1838, Cincinnati had become the nation's fastest-growing city, well-known as the "Queen City of the West." The time seemed ripe for Cincinnati to advertise its stellar performance through the mechanism of public exhibitions.[24]

The Mechanic's Institute directors projected the fair as the first of a series of annual events. Each would provide a formal, annual inventory of Cincinnati's material products "in the useful, ornamental, and fine arts." The fair offered all producers, whether independent craftsmen or large manufacturers, the occasion to demonstrate the quality of their goods to potential buyers. For an admission price of twenty-five cents, visitors gained entry to the three-day exhibition. Displays ranged from stoves to fabrics to bird cages. Judges awarded diplomas or certificates to the best items in each of a long list of categories on the basis of "fine finish" and "workmanship." Coupled with speakers on two evenings, the exhibition presented visitors "with much to admire, and something by way of instruction."[25]

On the last night of the fair, E. D. Mansfield, a prominent Cincinnatian and member of Daniel Drake's circle, summarized the significant elements of the event in his address on "The Worth of the Mechanic Arts." In Mansfield's view, "the simple fact that we can make such an exhibition, on the spot where it is made, at the time in which it is made, and surrounded by the community in which we live" demonstrated the maturity of the city of Cincinnati and of the mechanic arts in the incomparable American political and social environment. America, he said, was the most promising site for "that intellectual activity, which enters upon new enquiries, makes new discoveries, enlarges the mechanic arts, and extends the dominion of man over the kingdoms of nature." The American Constitution guaranteed the right to each citizen to think and act in any way not infringing on the rights of others, and the nation's progress had proved the Constitution's appropriateness. America was already taking its place among the civilized nations of the world, less than a century after its creation. Development of the "mechanic arts" by individuals of high moral character cooperating in the expansion of "that grand circle of knowledge—physical and spiritual—by which happiness is increased and humanity elevated" had played a prominent part in the nation's ascendancy. The mechanic arts, Mansfield said, were "united to all that pertains to social progress." It remained the genius of the American system that it allowed each citizen to contribute fully, ever widening the circle of knowledge.[26]

Bolstered by the exhibition's success, the Ohio Mechanic's Institute reaffirmed its commitment to making the fair a regular event. A "committee of arrangements" began early in 1839 to prepare the second annual fair. Items would be accepted for display only or for sale. Judges would award certificates "for such inventions as may be deemed of public utility; and also for excellence in workmanship." The judging criteria would reflect the committee's desire "to bring before the people specimens of the products of home industry [so] that both the merchant and the consumer may see at one view the variety and quality of our manufactures." The exhibition would become "the very best mode of advertising" Cincinnati's industrial prowess, noted one newspaper editor, as "merchants from abroad visit . . . to facilitate their selection of stock."

The fair itself did not disappoint. Categories represented included machinery, stoves, castings and gratings, mathematical instruments, edge tools, and agricultural implements, as well as harnesses, saddlery and trunks, carriages and carriage trimmings, chairs, and cabinet furniture and upholstery. Books and stationery, boots and shoes, chinaware, earthenware, copperware, hats, musical instruments, carpeting, sculpture, designs, and paintings were all enumerated, and provision was also made for "miscellaneous articles." Though the fair was planned to draw articles that were well made, its proponents maintained it displayed "such things as are daily made and offered for sale at the shops" of the city, not special exhibition models but "fair specimens of our best productions."[27]

Similar exhibitions followed in the years from 1840 through 1845. Each spring, the committee sent circulars soliciting participation from producers throughout the West, explaining that "the origin of these exhibitions is western, and their object is to bring into notice the products of western industry and skill; and to draw out and urge forward western talent and enterprise." Public exhibitions of this kind, according to the sponsor, advertised through "samples of what is regularly done" in the West the progress of the region, the importance of its urban center, Cincinnati, and the skill of its craftsmen. Manufacturer-mechanics agreed. While they refused to "resort to puffing contrary to the opinion of an enlightened age," they often referred to the institute's awards in advertising. The statement "FIRST PREMIUM BOOT—FOURTH ANNUAL FAIR" spoke for itself.[28]

The annual fairs marked the culmination of the program envisioned at the formation of the Ohio Mechanic's Institute. Its partisans credited it with encouraging "the achievements in mechanical science and art" currently "revolutionizing the face of the whole earth," and the variety and quality of goods displayed at the fairs illustrated its success. In the words of an institute commentator, such examples demonstrated that the institution had fostered in its participants "practical and scientific attainments, sobriety, and moral conduct." The program of "diffusing useful knowledge" through public lectures, classes, and library and reading room facilities had demonstrated its effectiveness in combating the problem of incivility in that part of America which was Cincinnati. It proved as well the conviction of its founders that the "duty and real interests of communities, as well as of individuals, are inseparably united." In Chevalier's words, "Cincinnati is in fact the great central mart" or "capital" of the West, "a large and beautiful town, charmingly situated."[29]

By the 1840s, after its first phase of existence, the program and emphasis of the Ohio Mechanic's Institute changed. Even the standard usage of the name shifted to "Mechanics' " and then simply "Mechanics." The institute did become committed later in the century specifically to training operative mechanics and training them technically rather than in broad general principles "useful" to all of society. Sometime after 1838, the institute began to respond to new problems in American civilization. Evidence of this transition to a new phase in the institute's operation can be found in the decreasing importance from 1840 on of public lectures and even of general scientific courses for members; in revision of the corporation's charter and constitution in 1846–47; and in the terms of the construction of a building in 1848 specifically for the institute.[30] New classes aimed to prepare their participants to practice specific operations, not to make rational observations of the world around them. No longer did the Ohio Mechanics Institute intend to serve a "mechanic" population so amorphous as to include virtually every member of the community. The changes made in the structure and operations of the institute in the 1840s and 1850s were effected to adapt it to serve newly perceived needs and circumstances so that it could survive into this new era in the nation's history.[31]

The alteration of the Mechanics Institute's corporate charter in the late 1840s exemplified the general shift in consciousness from concentration on the individual to focus on the group. In late 1846, a meeting of institute members, not directors, voted to petition the Ohio legislature for a change in their charter that would empower the whole membership rather than the directors alone to elect officers and to appoint committees. All directors resigned from the board in response, with one explaining that he was "disposed to give the members an opportunity of 'managing their own affairs' as they term it," agreeing with their contention that the members formed a group distinct from the directors. No longer did board members boast, as they had earlier, that any persevering individual could join their ranks and advance through self-education to the position of director. Instead, the resignation letter continued, the members "claimed that the Directors should be all *Mechanics* and that the *President* should be chosen from among the *active, energetic, young mechanics.*" If that was what "they" wanted, this director was willing to accept it, but not without commenting "that had not others, than operative mechanics, aided and assisted this institution, it would long ago been numbered with the 'things that were.' "

The first elected board under the new charter included only four of the previous sixteen directors, and the first president to be elected from the ranks was not Professor J. P. Foote, who had held the office since the institute's inception in 1828, but Miles Greenwood. As the owner of a steam engine and iron casting foundry, Greenwood, although not a typical operative mechanic, represented the rising manufacturing interest in the city, and heavy industry at that.[32] He soon became the principal contributor to the fund for what came to be called the "Greenwood Building," and he regularly submitted the names of his employees for institute membership. At least one commentator found a causal relationship between the success of the membership move for charter amendment and the construction of a new building specifically for the institute. While plans for a new institute building had been considered for six years, they came to fruition after the charter change, prompting one observer to call the charter the "leaven of true democratic impulse" that provided the impetus for construction.[33]

Notices in the papers calling for the support of "the Citizens of Cincinnati, and the mechanics of the city in PARTICULAR" point to the new orientation of the institute. Operative mechanics did respond to the call, for the institute's ledger of subscriptions included contributions as well as carpentry work, plastering, bricklaying, locks, and plumbing. Although some of these donations, especially those for larger amounts, came undoubtedly from proprietors offering the services of their employees, at least some workmen donated their own labor.[34] Speeches at the cornerstone-laying ceremony acknowledged the contributions of nonmechanic citizens and reminded the benefactors of the advantage they would gain in providing this setting to cultivate mechanics' "oppressed but aspiring minds." Training a class of "more intelligent mechanics than any other city in the world" offered Cincinnati the possibility of realizing its geographical potential to become the center of American industry.[35]

The institute program that emerged in the late 1840s differed sharply from that of its earlier years. The new institution was indeed dedicated to the training of operative mechanics. It also took a new place in the city, whose role itself had changed over the years. But in the first phase of institute activity, from 1828 into the transitional period around 1840, it functioned as a typical American response to the perceived problem of incivility in the new republic and served the whole population of the city, itself striving to take its place as a cultural as well as commercial center. "Diffusing useful knowledge" through the citizenry, the Ohio Mechanic's Institute served as a characteristic self-culture institution of the 1820s and 1830s.

Introduction to Chapter 2

The American Revolution's consequences, ideological and otherwise, were to resonate through American culture and society for decades to come. Although historians still argue over whether its character was democratic or not, it did have large consequences that did much to undermine the hierarchical culture and society of the colonial era, as in the movement toward separation of church and state, the freeing of slaves in the North, the spate of constitution-writing at federal and state levels of government, and the elimination of restrictions on suffrage in many states. Nor was this all. As Linda Kerber, among others, has pointed out, the Revolution even brought the issue of women's status in the new republic to the fore. Could the Declaration of Independence and patriarchy be reconciled?

And, indeed, the status of women as independent persons arose within the context of women's education, specifically concerning the purposes and content of a text of technical knowledge. In a prize-winning essay, "The American Career of Jane Marcet's *Conversations on Chemistry,* 1806–1853," M. Susan Lindee traces what she labels the American career of a European woman's textbook in chemistry. That book, Jane Marcet's *Conversations on Chemistry,* became the most successful elementary chemistry text in America before the mid-nineteenth century, going through multiple editions on this side of the Atlantic. The book was first published in London in 1806; its first American edition appeared that year as well. As Lindee points out, up to 1850 American publishers in five cities made twenty-three impressions of various editions of the book, and at least one contemporary estimated that it sold 160,000 copies in the United States. The book fared well in other respects too; a very imitative American text went through twelve printings in this country, and the book itself went through eighteen printings in Britain, four in Paris (and perhaps another, a plagiarist), and one in Geneva. In Germany an 1839 edition failed wretchedly.

Lindee insists that Marcet's was the most widely used chemistry text in women's schools and academies in America. Yet Jane Haldimand was no icon of democratic culture. Born in 1769 in London into a prosperous Swiss family, when she was thirty she married Alexander Marcet, a London physician and chemist; their social circle included such illuminati as Harriet Martineau, Thomas Malthus, J. J. Berzelius, and Humphry Davy. Her top drawer social connections facilitated a long writing career for this genteel, if not precisely titled, lady, for she wrote similar books on political economy in 1817, natural philosophy in 1820, and vegetable physiology in 1829, all with great successes as books.

And here seemed to be a perfect example of the processes by which a lady of standing taught other aristocratic ladies, and even democratic women in an emerging egalitarian culture, the underlying principles and theories of a body of knowledge on a par with that of men, in the first half of the nineteenth century. Jane Marcet's *Conversations on Chemistry* was no inducement to feminine submissiveness. No matter what changes the book's American editors and publishers made in it, the text nevertheless did not fulfill the essentially conservative function of popular science for women's domestic needs and spiritual uplift that publishers apparently intended. Rather it was a radical tome for its times. It was a subversive text that taught women what was for its day technical knowledge, in this instance technical knowledge in chemistry, that was usually reserved for men, meaning the intricacies of contemporary chemical theory and hands-on laboratory instruction in the field. The tome's subversiveness resided precisely in its intended feminine audience, which was assumed to be just as worthy to receive said technical knowledge as any man. Lindee insists that the popularity of Marcet's book implies that American educators as well as editors and publishers wanted female as well as male students to grasp the fundamentals of theoretical and experimental science. This kind of instruction only made sense if its purpose was to prepare young women to be teachers, not mere domestic creatures. In part this was the consequence of the cultural shift from the aristocratic lady of the eighteenth to the democratic woman of the early nineteenth century, not to mention the individualism of that latter era, in which the new republican society was constituted of civilized individuals who pursued commerce and trade without interference from government or other aristocracies or monopolies. But Lindee's essay reveals a new cultural and social role for the individual or democratic woman of the early nine-teenth century, just as Spraul-Schmidt's lays bare the importance of the individual more generally. Thus in this and other ways did larger notions of the individual come to shape and inform the fortunes of technical knowl-edge and its purveyors.

M. SUSAN LINDEE

2

The American Career of Jane Marcet's *Conversations on Chemistry,* 1806–1853

Jane Haldimand Marcet's *Conversations on Chemistry* has traditionally claimed historical attention for its effect on the young bookbinder Michael Faraday, who was converted to a life of science while binding and reading it. Marcet "inspired Faraday with a love of science and blazed for him that road in chemical and physical experimentation which led to such marvelous results," in H. J. Mozans's romantic account. Or, as Eva Armstrong put it, Marcet led Faraday to "dedicate himself to a science in which his name became immortal."[1]

In these accounts Marcet is important for her effect on one prominent male scientist. But her influence was much wider: *Conversations on Chemistry* was the most successful elementary chemistry text of the period in America. American publishers printed twenty-three editions of Marcet's text, and twelve editions of an imitative text derived from it. Many young men and women had their first serious exposure to chemistry through the lively discussions of Mrs. B., Emily, and Caroline, the characters Marcet used to convey her ideas. The book was widely used in the new women's seminaries after 1818. There is also evidence that young men attending mechanics' institutes used Marcet's text, and medical apprentices favored it in beginning their study of chemistry.[2]

The widespread use of Marcet's book in the early women's schools is of particular interest. Allusions to domestic applications of science and the spiritual insight it offered were commonly used to justify science instruction for women in these new institutions. But did the texts and style of instruction bear out that justification? If they did not, what might this suggest about the goals and intentions of those offering scientific training to young women?

I have compared Marcet's *Conversations* with other elementary chemistry texts published in the United States between 1806 and 1853. My purpose is to shed some light on the priorities of a poorly understood group: teachers and administrators at intermediate or college-level women's schools in the first half of the nineteenth century. I show that while these educational reformers had numerous options, they favored a chemistry text that was theoretical and experimental: Marcet's *Conversations on Chemistry*. More "domestic" or practical chemistry textbooks, which were widely available, fared poorly, as did the less common textbooks emphasizing chemistry's spiritual lessons.

School administrators and instructors used domestic and religious justifications to increase the social acceptability of science education for women in the early nineteenth century. My work suggests, however, that the actual instruction at the women's schools promoted feminine interest in scientific theory at a level that exceeded that required for domestic efficiency or religious gratification.

Jane Haldimand Marcet (1769–1858) was born in London of a prosperous Swiss family. When she was thirty years old, she married Alexander Marcet, a London physician and chemist. Her husband's social circle included J. J. Berzelius, Humphry Davy, the botanist Augustin de Candolle, the mathematician H. B. de Saussure, the writers Harriet Martineau and Maria Edgeworth, the political economist Thomas Malthus, the physicist and naturalist Auguste de la Rive, and the chemists Pierre Prevost and Marc Auguste Pictet.[3]

Such social connections gave Marcet access to new ideas, and she translated this access into a long, productive writing career. After the success of *Conversations on Chemistry,* her first book, came *Conversations on Political Economy* (1817), *Conversations on Natural Philosophy* (1820), and *Conversations on Vegetable Physiology* (1829). She published anonymously until 1837, and for this and other reasons her works were often attributed either to other women writers or (in America) to the male commentators whose names appeared on the title page. Marcet's use of "Mrs. B." as the instructor in these conversations led to speculation that the author was Margaret Bryan, a British popularizer of science already prominent when Marcet was a child. Marcet may, of course, have chosen "Mrs. B." as an allusion to Bryan. Marcet's *Conversations* was also attributed to other women writing about science, including Sarah Mary Fitton, who wrote *Conversations on Botany* in 1817.[4]

All of Marcet's later *Conversations* involved the characters—Mrs. B., Caroline, and Emily—introduced in *Conversations on Chemistry*. Caroline, an impetuous and skeptical student, was somewhat more interested in explosions than in fundamentals of science. Emily was serious and bright and more likely to ask important questions. The two young women were thirteen to fifteen years old (at least, Emily's age was given in *Conversations on Natural Philosophy* as thirteen) and apparently not related. Caroline's father owned a lead mine in Yorkshire, and Emily's family background was not mentioned.[5]

They were young women of wealth, well educated and sensitive to social conventions. In her introduction to the chemistry text Marcet apologized for their intelligence: "It will no doubt be observed that in the course of these Conversations, remarks are often introduced, which appear much too acute for the young pupils, by whom they are supposed to be made. Of this fault the author is fully aware." She explained that the unusual brightness of the pupils was necessary lest the work become "tedious."

In the opening conversation, Caroline claimed to be uninterested in the science of chemistry:

Caroline. To confess the truth, Mrs. B., I am not disposed to form a very favourable idea of chemistry, nor do I expect to derive much entertainment from it. I prefer the sciences which exhibit nature on a grand scale, to those that are confined to the minutiae of petty details.

Mrs. B. I rather imagine, my dear Caroline, that your want of taste for chemistry proceeds from the very limited idea you entertain of its object. . . . [Nature's laboratory] is the Universe, and there she is incessantly employed in chemical operations. You are surprised, Caroline; but I assure you that the most wonderful and the most interesting phenomena of nature are almost all of them produced by chemical powers.

When the conversation turned serious, Emily joined in, and the first lesson centered on "constituent" and "integrant" parts. The book then progressed in twenty-six conversations from simple to compound bodies, and from elements to living systems. Marcet included discussions of light and heat, electricity, oxygen and hydrogen, sulfur and carbon, metals, attraction, acidification, decomposition, and animal productions. A twenty-seventh conversation, on the steam engine, was added from 1830 on.

This range of topics indicates the parameters of early nineteenth-century chemistry. The field included—in some fashion—geology, mineralogy, electricity, fermentation, plant respiration, and animal growth. Chemists studied meteors, minerals, animal phosphorescence, medicinal cures, and soil samples. Marcet could, quite reasonably, turn her attention to "bones, teeth, horns, ligaments and cartilage" or to the "effects of Light and air on Vegetation."

But chemistry, however broadly defined, was not the only subject of the text. By the time Marcet wrote her chemistry book, she had already completed her *Conversations on Political Economy* (published later), and some of the themes from that volume made their way into her chemistry lessons. For example, she touched on problems of class. She had Mrs. B. proclaim that the "well-informed" were often too eager to adopt new technology, while the uninformed, "having no other test of the value of a novelty but time and experience," were sometimes able to "prevent the propagation of error." Mrs. B. also praised England's colliers, "digging out of the bowels of the earth one of the most valuable necessaries of life." She expressed disdain for scientific pretense, urging Caroline not to use the word *oxydate* rather than *rust,* "for you might be suspected of affectation."[6]

Marcet also was aware of the sexual politics of her work and made frequent reference to her feminine readers and their presumed interest in science. In her preface she apologized for daring to publish a work on science, describing her apprehension that her work would be considered "unsuited to the ordinary pursuits of her sex." But the recent establishment of public institutions "open to both sexes, for the dissemination of philosophical knowledge" proved that "general opinion no longer excludes woman from an acquaintance with the elements of science."

She explained that her interest in chemistry was aroused by attendance at the public lectures of Humphry Davy, which she initially found confusing. When the basic concepts of the new chemistry were explained to her in "familiar conversations," Marcet said, she could enjoy Davy's lectures much more. "Hence it was natural to infer that familiar conversation was, in studies of this kind, a most useful auxiliary source of information, and more especially to the female sex, whose education is seldom calculated to prepare their minds for abstract ideas, or scientific language." Her book was written because "there are but few women who have access" to scientific friends, such as her own, willing to converse with them about theory. (John Crellin has suggested that Marcet's "scientific friend" was almost certainly her husband.)[7]

At the same time, she did not promote an unseemly female participation in science. When Caroline mentioned pharmaceutical chemistry, Mrs. B. proclaimed that pharmaceutical work "belongs exclusively to professional men, and is therefore the last [branch of chemistry] that I should advise you to pursue."[8]

In its approach to chemistry, Marcet's book was theoretical rather than practical. She updated her treatment of important ideas in later editions, and at least on some topics her elementary text kept pace with scientific changes. She followed Antoine-Laurent Lavoisier's scheme of classification of the elements, as laid out in the 1796 English translation of his *Traité élémentaire de chimie,* considering light, electricity, and caloric "imponderable agents." She somewhat conservatively clung to the caloric theory, however, even after

Davy had abandoned it. She used a Newtonian, corpuscular theory of matter, and she explained chemical reactions in terms of affinity, aggregation, gravitation, and repulsion.[9]

Marcet did not mention John Dalton's atomic theory until after 1819, and even then she expressed doubts about its validity. This in part reflected Davy's skepticism. But it was a skepticism widely shared; many other writers of chemistry texts, including Thomas Thomson, W. T. Brande, and Andrew Ure, continued to question Dalton's theory (as an explanation for the fundamental nature of matter) until as late as 1841.[10]

Despite its elementary nature, Marcet's treatment of chemical theory compared favorably with that of the Scottish chemist Edward Turner in his much-admired college-level textbook *Elements of Chemistry*. Turner's book, first published in 1827, was about as widely used and imitated as Marcet's. He had at least three near-plagiarists in America: Lewis C. Beck, John Lee Comstock (one of Marcet's editors), and John Johnston all produced chemistry texts that depended heavily on Turner. They borrowed his organizational format, illustrations, charts, appendixes, and in many cases his words. All acknowledged their debt to Turner, stating that their work was "on the basis of Turner's *Elements of Chemistry*" or that Turner was "used more freely than any other" author.[11] Both Turner's text and those drawn from it were popular in American men's colleges.

Marcet's handling of chemical theory was remarkably consistent with Turner's. In a point-by-point comparison of the two authors' treatment of heat as an "imponderable substance," the correlation of the subjects explored and scientists cited is very high. Both discussed William Herschel's studies of light and heat, John Leslie's work with the radiation of heat, Marc-Auguste Pictet's *Essai sur le feu,* Count Rumford's work on clothing and the conduction of heat, William Wells's theory of the formation of dew, the problem of cold as the absence of heat (rather than as a negative quality), the use of a pyrometer, and Pierre Prevost's studies of radiation.[12]

Marcet's text also kept pace with William Brande's *Manual of Chemistry*. Brande was professor of chemistry at the Royal Institution (appointed to replace Humphry Davy in 1812), and his text was intended as an advanced accompaniment to his three-month lecture course for men, which he taught with Faraday as his assistant. Brande's and Marcet's classification schemes were similar, with Marcet's in some ways superior. She organized the elements on the basis of their presumed nature, and she attempted to construct a meaningful system that would help her students understand the processes of chemistry. Marcet's book would have covered Brande's course adequately, for the topics listed in his syllabus and those discussed by Marcet were largely the same.[13]

These comparisons suggest that Marcet's book was no collection of tips for homemakers and farmers, but an introduction to the most important chemical theories of her day. Its popularity in the new women's schools in America therefore raises questions about the goals and priorities of the educational reformers who taught there.

Conversations on Chemistry was first published in London in 1806. The first American edition appeared later that same year. From 1806 to 1850, American publishers made twenty-three impressions of various editions of the work, at Hartford, Boston, Philadelphia, New Haven, and New York.[14] There were also twelve printings of a highly imitative American text, *New Conversations on Chemistry*, by Thomas Jones. Marcet's book was almost as popular in Britain, going through eighteen printings.[15] It was printed four times in Paris (perhaps more, since Marcet had at least one French plagiarist) and once in Geneva.[16] The book was a failure in Germany, where a single 1839 edition sold poorly.[17] A contemporary commentator set American sales figures at 160,000 copies.[18]

Marcet did not intend her *Conversations* to be used as a textbook. In Britain, it was apparently used as she expected, as a guide to popular lectures on chemistry or natural philosophy. But in the United States it became the most successful elementary chemistry text of the first half of the nineteenth century. A succession of male editors reshaped it for classroom use through twenty-three pirated American editions over forty-seven years. Indeed, as noted earlier, the work was commonly attributed (in biographical dictionaries, catalogues, and obituaries in the United States) to its male editors.[19] In the absence of international copyright law, Marcet received no income from these American editions, nor had she any control over the American commentaries and improvements.[20]

The American editors added study questions, dictionaries of terms, guides to the experiments, and critical commentaries. These amendments for the classroom were not a marketing strategy concocted by the books' U.S. publishers, but the response of professional chemists and educators to the books' growing use as an introductory chemistry text. *Conversations on Chemistry* was widely adopted in the schools by 1818. It then attracted American editors, most of whom seemed to be disturbed by its popularity.

Marcet's American commentators included a minister and four professors of chemistry or chemical lecturers. They worried about questionable theories (Davy's) and dangerous experiments. They also attempted to promote American scientists—Robert Hare, whom Marcet neglected, and Benjamin Franklin, whom she misinterpreted.[21]

Marcet's most frequent editor was John Lee Comstock, who made his debut anonymously in the fourth edition of 1818 as "an American gentleman." His name first appeared in the 1822 Hartford edition. In 1826 O. D. Cooke produced an edition of *Conversations on Chemistry* with both Comstock's commentary and a series of numbered "study questions" provided by the Reverend John Lauris Blake. The combination of Comstock's criticisms and Blake's questions was the standard format for most American editions throughout the rest of the book's career.

Blake (1788–1857) was an Episcopalian minister in Boston. He had resigned his rectorship in 1822 to devote himself to "literary work," which included writing an introductory astronomy book and providing numbered study questions to both Marcet's *Conversations on Chemistry* and her later *Conversations on Natural Philosophy.* Blake must have been interested in the education of women: he started a girls' school at Concord, New Hampshire.[22] His questions in Marcet's chemistry book (1,456 of them in the 1836 Hartford edition) were printed at the bottom of each page and intended to aid in classroom instruction. On the title page Blake warned (in triple negative) that "no small portion of learners will pass over without study, all in which they are not to be questioned."

Blake's questions were not particularly thought provoking—they promoted rote learning—but they were apparently taken seriously by some students. In several copies of Marcet's text reviewed for this study, some long-ago student had dutifully penciled in the proper answers to these questions in the small space allotted on the page.

Comstock (1789–1858) was a self-educated surgeon who served in the army in the War of 1812 and later settled in Hartford to write and edit textbooks on chemistry, natural history, botany, physiology, and mineralogy. Comstock's "original" work was apparently often borrowed from European authors. The *Dictionary of American Biography* credits him with the authorship of a *History of Gold and Silver,* a *History of the Greek Revolution,* and a *Cabinet of Curiousities.*[23] He also wrote a highly derivative chemistry text: his *Elements of Chemistry* (1831) was a much-simplified and quite popular version of Turner's text of the same name. His 1822 *Grammar of Chemistry* was apparently also borrowed from another author. It was written "on the plan of David Blair," a pseudonym of R. Phillips, and "adapted to the use of schools and private students by familiar illustrations and easy experiments."[24] And Comstock's *Conversations on Natural Philosophy* was in fact Marcet's work, with his name on the title page as editor.

Two other American editors, the Philadelphia chemistry professors William H. Keating and Thomas Cooper (who produced one edition each), merely inserted a few mild footnotes clarifying Marcet's experiments or ideas.[25] But Comstock, her first and most persistent American editor, provided from 156 to 173 notes in his various editions, for a text averaging about 330 pages. In these notes he frequently disagreed with Marcet and sometimes implied that she was incompetent. When she explained the presence of so much "calcareous matter" as the "effect of a general combustion occasioned by some revolution of our globe," Comstock noted: "This idea is at random. We cannot account for the origin of carbonic acid in its native state, any better than we can for oxygen."[26] When Marcet suggested that it was highly unusual for three or more substances to combine without any of them being precipitated, Comstock noted that "such compounds are quite numerous." He characterized her explanation of volcanoes as "supposition piled on supposition."[27] When she attempted to explain the role of water in the life cycle of plants, he responded in a footnote: "The foregoing paragraph might mislead the student. Indeed, it seems to have been written without regard to proper authorities." When she suggested that "combustion is the result of intense chemical action," he responded: " 'Intense chemical action' neither explains the process, nor, indeed conveys to the mind any definite idea."[28] And when she said the concepts of negative and positive indicated "different quantities of the same kind of electricity," Comstock replied (with italics): "In this chapter, Mrs. B. has used these terms of the American philosopher [Franklin] improperly, for *plus and minus* were never meant to signify two sorts of electricity, but only its *presence or absence.*"[29]

If Comstock disapproved of many of Marcet's proposals (both theoretical and experimental), why did he continue to edit the book vigorously for four decades? His introduction provides a partial explanation: Comstock was worried about the book's widespread use in the classroom. "Known and allowed facts are always of much higher consequence than theoretical opinions," he said in the "Advertisement of the American Editor" that introduced his editions. "A book designed for the instruction of youth, ought, if possible, to contain none but established principles."[30]

Keating and Cooper, while milder in their criticisms of Marcet, also expressed concern about the promotion of questionable theories to beginning students. Cooper edited the text "lest the young student should adopt as certainties many theoretical views which have hardly yet arrived at probability." He noted that Marcet had followed Davy where his contemporaries "have not yet dared to follow him." This adoption of Davy's ideas rendered the book "extremely interesting" but less than ideal for instruction in the fundamentals of chemistry.[31]

Marcet's editors also worried about her depiction of the use of hands-on laboratory experiment in the training of beginners. They found such a proposal extremely risky, and their concerns were not unwarranted. From Comstock's corrections of her experiments, it appears possible that Marcet did not actually perform all the experiments she described. Certainly her suggestion that elementary chemical instruction might include laboratory experiment was quite novel. Indeed, in 1822 her editor William Keating, of the University of Pennsylvania and the Franklin Institute, was one of the first to apply this teaching method in an American college. It was not until after the Civil War that laboratory instruction for beginning students became the norm.[32]

While her other American editors merely inserted footnotes or study questions, Thomas Jones wrote a new text that followed Marcet's format precisely in terms of data presented, but eliminated the humor and personal commentary of the original. Jones, a professor of chemistry at Columbia College in Washington and a popular lecturer on chemistry and natural philosophy, was interested in filling the text with as many chemical facts as possible.[33] Publishing his first version of Marcet's book in 1831, he explained that while Marcet's text received deservedly high praise and had contributed more than any other work to promote the study of chemistry, its original role as companion for the parlor had been superseded. The new role of textbook called for a different presentation. The digressions that gave the original work variety and interest in the family circle were now an impediment to the rapid assimilation of new facts, he said.[34] Jones's version, though lacking the entertainment value and charm that might be assumed to be one reason for Marcet's success, was relatively successful itself: it was reprinted twelve times, more frequently than most other chemistry texts of the era.

Marcet's American editors suggested that she went too far in her promotion of the latest chemical theories. Yet her discussions of theory may have been what academy-level instructors found so attractive. And the proposed experiments her editors found so risky may have made her work more valuable to instructors hoping to spark young women's interest in science.

The antebellum women's academies have been a subject of increasing historical interest since 1979. Science instruction at these institutions was touched on in Thomas Woody's classic 1929 history of education for women. In 1979 Deborah Jean Warner examined more precisely the kinds of instructions and instructional materials that women's academies offered. Linda Kerber and Anne Firor Scott have explored their complex cultural role, suggesting that practice was not always in line with public rhetoric. Those promoting women's education for the sake of republican motherhood (the rearing of good male citizens who could defend the republic) may have had more radical intentions. And as Patricia Cline Cohen has shown, women's education in mathematics was predicated on the household applications of numerical reasoning (as in knitting or cooking), while actual instruction was much more advanced than these simple tasks required.[35]

Certainly the historical picture of both the women's academies and the role of science therein is incomplete. Some sciences, including chemistry, were more widely taught in the women's academies than in boys' high schools of the early nineteenth century. And at least some women's schools, particularly Emma Willard's Troy Female Seminary, offered a greater range of sciences than did contemporary men's colleges.[36] Laboratories and observatories at the female colleges were not well funded, but they represented the single largest investment, excepting buildings, at many schools. A women's school in New York City, Abbott Collegiate Institute, claimed scientific apparatus unsurpassed in character by that of any other institution in the country. Astronomical equipment was particularly popular. Albany Female Academy and Packer Collegiate Institute each owned an orrery, a moving, mechanical representation of the solar system, made by a renowned Kentucky instrument maker.[37]

Such equipment, as Deborah Jean Warner has noted, proves nothing about the quality of science teaching. The paraphernalia was as important for promotional as for educational reasons. Yet she argues that other evidence suggests that the quality of the instruction in some sciences was relatively high. Some lecturers appearing at the women's schools were well known (Benjamin Silliman, Jr.; Elias Loomis), and some science teachers were extremely competent, among them Alonzo Gray, who taught at Brooklyn Female Academy, and Louis Agassiz, who with his wife Elizabeth ran a school for girls in Cambridge from 1855 to 1863.[38]

The availability of scientific apparatus and the high quality of some instructors suggest that science education at the women's academies was more than a public relations ploy. The selection of textbooks reinforces this conclusion. Those teaching chemistry to young women in this period had numerous options. Their choice of Marcet's text indicates their educational priorities. It suggests that their commitment to scientific instruction for women was not completely encompassed in their publicly stated goals. Textbooks conforming more properly to these stated goals were widely available before 1840. Most emphasized the practical applications of chemistry. But at least one important American chemistry text focused on the spiritual lessons it provided. This was the text of the American educator Almira Hart Lincoln Phelps, the sister of Emma Willard.[39]

Phelps should have had considerable insight into the instructional materials needed in the new women's schools. Yet her academy-level chemistry text, specifically intended for the instruction of young women, was a failure; it was reviewed unenthusiastically and printed only twice, in 1838 and 1842.[40] Phelps's error may have been her assumption that chemical education was a form of religious instruction. While Marcet mentioned the relevance of chemical theory to religious faith only casually, Phelps's *Familiar*

Lectures on Chemistry was metaphysical throughout. She said chemistry could provide lessons in humility—our own bodies are composed of a few elements of the same nature as those that form the very worm that crawls—and in hubris: there is a portion of ourselves that is beyond the scope of chemical science, which cannot be analyzed, because it is incapable of being separated into parts.[41]

While Phelps's primary interest was in the spiritual lessons of science, she also recognized that chemistry had a peculiarly practical aspect. She assigned her pupils to explain the chemical principles involved in making bread and informed them that chemistry had an important relation to housekeeping in the making of gravies, soups, jellies and preserves, bread, butter and cheese; in the washing of clothes; in the making of soap; and in the economy of heat in cooking and in warming rooms.[42]

Other writers considered the utilitarian aspect of chemistry its chief value to potential students. The useful purposes these writers selected for discussion shifted with the intended audience. An author intending to address the problems of household science might discuss the relevance of chemical facts to the fermentation of bread, preservation of milk and butter, sources of impure air in the home, and properties of fuel used for artificial heating. Another, intending to reach workingmen, would focus on leather tanning, wine brewing, soil analysis, and medicine. John R. Coxe's translation of M. J. B. Orfila's *Practical Chemist* (1818) contained little chemical theory, focusing instead on information useful to the pharmacist, farmer, or physician. Similarly, William Henry's *Elements of Experimental Chemistry* classified metals practically, rather than theoretically, and dealt solely with the relation of chemistry to the practical arts. The American physician and Harvard chemistry professor John Gorham deemed even Henry's chemical text too experimental, and in his *Elements of Chemical Science* (the first original American chemistry textbook) simplified Henry's approach by eliminating virtually all laboratory work. John Lee Comstock's own text, *Elements of Chemistry,* first published in 1831, was an entirely descriptive and practical text that gave no attention to chemical theory. Even as late as 1867, J. Dorman Steele's popular *Fourteen Weeks in Chemistry* concerned only that practical part of chemical knowledge necessary in the schoolroom, kitchen, farm, and shop. And a masculine version of Marcet's *Conversations on Chemistry,* the Reverend Jeremiah Joyce's *Dialogues in Chemistry,* featured conversations between a tutor and two male pupils, Charles and James, on the relevance of chemistry to agriculture, gardening, and the arts of cooking and of making wine, beer, and other fermented liquors.[43]

An introductory text that combined all these interests rather broadly was produced in 1822 by the New York educator Amos Eaton, a friend of both

Almira Hart Lincoln Phelps and her sister Emma Willard. Eaton dedicated his *Chemical Instructor* to Willard because she was the first in the interior of the northern states to introduce experimental chemistry into (public) schools.[44] Eaton's text, written to replace Marcet's, which he disliked, was intended for the audience—academy chemistry instructors—who had already demonstrated their enthusiasm for her approach. Eaton interpreted that market as receptive to a practical treatment of the subject. He was unwilling to let a single chemical idea or principle pass without mentioning a practical application: he made special appeals to those engaged in the full-time management of a house. His intentions were egalitarian and democratic. He proposed simple, inexpensive experiments, recognizing that his readers might not have access to expensive chemical equipment or rare materials; part of his objection to Marcet was that she assumed her readers would have ample access to equipment and supplies.[45]

But Eaton's book was not widely used in the women's academies. Instead, many instructors of young women continued to introduce chemistry through Marcet's *Conversations on Chemistry,* a work that overlooked the domestic or practical applications Eaton and other American writers believed to be so important.[46]

By the 1820s popular science tailored to a female audience was a well-accepted social activity. From these public lectures and popular books women supposedly gained lessons in piety and useful household tips. School administrators at the women's academies transferred this reasoning to the formal educational setting. They offered their students those sciences promoted for women in popular lectures and books: natural philosophy, astronomy, chemistry, and botany.

But popular lectures and popular science books were casual entertainment, essentially conservative, legitimated by the presumed domestic and religious applications of scientific knowledge. Education at the female academies entailed institutional approval of a sustained course of study of science, however elementary; there was often an implicit expectation that some students would pursue careers as teachers. While both activities were justified in similar ways, they reflected fundamentally different assumptions about female involvement in science. The conservative arguments that made sense of science education for women apparently had little impact on actual scientific instruction, which (at least in the case of chemistry) was often focused less on spiritual or domestic applications than on chemical theory and experiment.

Despite competition from dozens of other texts, Jane Marcet's *Conversations on Chemistry* dominated elementary chemical instruction in these academies. Administrators could have chosen texts that emphasized useful applications or spiritual lessons. They chose instead a presentation novel for both its attention to chemical theory and its advocacy of hands-on laboratory instruction for beginners. It was not simply a matter of teaching the principles of baking or soap making. Academy chemistry, at least in those schools that used Marcet's text, was serious chemistry for beginners: an up-to-date review of European chemical theory, illustrated by experiment, requiring an understanding of chemical terminology and facility in the manipulation of laboratory equipment and chemicals.[47]

The popularity of Marcet's book suggests that American educators wanted young women to understand the basics of theoretical and experimental science. Their reasons for this remain unclear. But certainly the instruction offered in the women's academies provided an important initial impetus for changes in the nature of women's participation in science. While the legacy of scientific training in the women's academies is difficult to measure, some women did become prominent scientists in the second half of the century. Wellesley College's first professor of physics, Sarah Frances Whiting, graduated from Ingham University for Women and taught at the Brooklyn Heights Seminary. The naturalist Lydia White Shattuck studied at Mount Holyoke Seminary. The botanist Graceanna White Lewis attended the Kimberton Boarding School. The astronomer Maria Mitchell, her student and fellow astronomer Mary Whitney, the chemist and educator Mary Lyon, the psychologist Christine Ladd-Franklin, and the chemist and home economist Ellen Swallow Richards were also products of this changing educational climate.[48]

The availability of serious scientific education in the new women's academies set the stage for increasing women's involvement in science. The access to introductory science instruction in a formal laboratory setting—rather than through a male family member or a brother's tutor—legitimated feminine interest in scientific theory. And as the famous Faraday anecdote suggests, the young mind can sometimes reach grand conclusions from rather minor encounters.

Introduction to Chapter 3

How the larger culture helps shape both technical knowledge and the technicians' notions of their own identity is suggested in Alan I Marcus's essay on the shift from individual practitioner to regular physician in mid-nineteenth-century Cincinnati. Taking the history of Cincinnati medical societies as his narrative or empirical base for understanding the larger culture, Marcus probes the arcane histories of several antebellum Cincinnati medical societies. In particular Marcus finds a large cultural shift, from an age venerating the individual to one stressing the reality of group identity and membership. Thus in the early nineteenth century—to the late 1830s, that is—medical men followed a code of conduct of genteel individualism, a kind of Jeffersonian or republican elite of talent and merit in which manners, morals, and expertise were juxtaposed in the same individuals and in which communal association with one's fellows kept a doctor's behavior on the up and up. Marcus insists that the early nineteenth-century medical societies and the medical knowledge that they preserved were fully a construction of contemporary society and culture, in which the individualistic perspective reigned supreme. As each individual doctor was a free agent, he could better himself. Success was possible. It was measured in terms of what one did, rather than what one had, such as inherited status and wealth. This gave the licensed doctor an enormous competitive edge over his unlicensed colleague.

Starting in the 1830s, however, Marcus insists, doctors began to define themselves in new kinds of ways and thus created a very different nexus of cultural notions, institutions, and technical discourses for the practice of medicine in American society and culture. According to this new definition,

53

or redefinition, doctors were not mere individuals. They belonged to distinct and different groups of medical practitioners. Each group of doctors had their own intellectual system of medicine, their own *materia medica*—their own medical systems. Medical systems were not codifications of genteel individualism. Rather they were sets of communally shared beliefs, practices, and behaviors. Membership in a group and identification with its beliefs and practices—these were what mattered now. Thus Marcus lays bare the key to the seemingly undecipherable warring of medical sects in post-1830s America: the invention of the group rather than the individual as the key to society and culture. Technical knowledge—in this instance, knowledge about doctors, their diagnoses, their remedies, and the like—underwent dramatic shifts and transformations, in which old knowledge was reshaped into new knowledge, seemingly as the result of a shift in the culture's larger notions of how society and nature were constructed.

From Individual Practitioner to Regular Physician: Cincinnati Medical Societies and the Problem of Definition among Mid-Nineteenth-Century Americans

3

The Medico-Chirurgical Society of Cincinnati and the Cincinnati Medical Society never fulfilled their founders' expectations. Sparsely attended, infrequently held meetings with programs members treated with indifference continually plagued the two organizations. This lack of enthusiasm indicates that to the vast majority of Cincinnati doctors, neither the Medico-Chirurgical Society, founded in 1848, nor the Medical Society, formed in 1851, seemed a useful or appropriate forum. In 1857, a group from the Medico-Chirurgical Society proposed and established a new medical society, the Academy of Medicine of Cincinnati. The new society differed in structure from its predecessors. Earlier medical organizations had chosen their members quite carefully and conducted sessions in camera. The academy, on the other hand, welcomed all regular medical practitioners, threw its doors open to medical students, and regularly published its transactions in city medical journals. The academy's objectives also differed strikingly from those of earlier area medical societies. According to its first president, the venerable Reuben D. Mussey, the academy sought to make "its proceedings the basis of public opinion in matters pertaining to medicine."[1]

The radical departure by the Academy of Medicine from established medical society norms suggests that it was grounded on new premises. Emphasis on inclusiveness rather than selectivity, for example, implied that its founders assumed a certain homogeneity among regular physicians. Its desire to mold public opinion rather than restrict deliberations to members suggested that these founders understood that the group's status depended on public approval. Finally, their decision to publish its transactions in medical periodicals—a frank attempt to reach out to regulars unable or unwilling to attend its meetings—betrayed their realization that each regular physician's status and well-being were bound up with those of every other. Actions of one could discredit them all, or so they believed. Put more simply, the academy's founders identified a specific problem and proposed that the academy, established as a different kind of medical society than its predecessors, would provide the remedy.

The Cincinnati doctors' rejection of their previous institutional forms and abandonment of precedent did not mark them as unique. The collapse of established American institutions in the years after 1840 and their replacement with new, quite different bodies was a pervasive antebellum phenomenon but its appearance is noted by scholars only *en passant*. Several have detected this event within their own narrow case studies. For example, they have uncovered a new mid-century technical education, a significant reorganization of the temperance movement, and a new meaning of manifest destiny. Others have focused on the formation of new national political parties, the change in the pursuit of science and establishment of new scientific institutions, the unfolding of new economic theory and development of a nationwide economy, large-scale introduction of new production techniques, restructuring of American Protestantism, and emergence of a new type of Judaism, Reform Judaism. I have elsewhere discussed creation of year-round, regular municipal social service agencies. But by viewing these occurrences as isolated, as peculiar to different areas of examination and investigation, scholars have failed to recognize their deep cultural penetration and significance. When taken together, these limited studies suggest that institutional reformulation constituted a fundamental, if unacknowledged, facet of antebellum culture, striking at the very fabric of American society.[2]

This mid-nineteenth-century institutional revitalization was a matter of definition. The massive nationwide rejection of extant institutions and formation of new ones reflected a new perception of social reality, a new mid-nineteenth-century notion of American civilization. Americans moved forcefully to put their institutions in accordance with their new notions. The Academy of Medicine of Cincinnati bears legacy to that dramatic act. Its creation provides a convenient microcosm through which to view the pervasive mid-nineteenth-century national reexamination. The issues surrounding the academy's formation were characteristic questions in the United States after about 1840.

Prior to the academy's creation, city medical societies were voluntary associations of individual practitioners. There was an element of class here. Cincinnati's most esteemed medical men began them. These city societies catered only to those gentlemen who had demonstrated that they possessed the proper character, the manners and temperament, for medical practice in polite society. Often equipped with libraries and cabinets, these organizations of civilized medicine-practicing men empowered each member to peruse the medical texts and specimens. They also fulfilled their members' desire to come into regular contact with similarly interested, high-minded individuals. These medical society attributes enabled each gentlemen–physician member to use the organization to improve himself further. Calling these bodies "societies of emulation," Daniel Drake, Cincinnati medicine's high priest, noted that "the physician, however learned, needs the influence of professional associations to repress self-sufficiency, avert mannerisms and to arouse his ambition." These associations, argued Drake, led to "improvement in thinking and speaking that always result from debate, from the juxtaposition of inquisitive minds and the collisions that are the natural result of such a situation."[3]

Membership in these early medical societies brought other advantages. It was tantamount to certification, which usually translated into substantial numbers of paying patients as well as into public standing. Medical society membership testified to an otherwise ignorant public that physician members possessed an exquisite sense of ethics, had mastered the social graces, and were acquainted with medical science. A mannered practitioner deserved to lead the community. Medical association members were characterized by "justice, morality and benevolence." Avoiding "all appearances of outward show and idle parade," medical gentlemen rejected "boastful arrogance, proud assumptions of superiority, unscientific vaporing and unscrupulous cupidity." Conduct deviating from this proper, implicit code seemed despicable. Drake termed it "unphilosophical—ungentlemanly—and, therefore, unprofessional."[4]

That philosophy also guided society members in their medical science pursuits. They claimed that their methodology, like their manners and morals, set them apart. "Idle speculation" and "vain theorizing" held no place in civilized medical practice. "Experimental investigation," the compiling and ordering of data without a priori assumptions, marked the civilized doctor. All diseases, Drake reminded his medical colleagues, "should be diligently and thoroughly studied: that the modifications which they exhibit in successive years and in different situations should be compared with each other as well as the various means of cure which have been found most successful."[5]

Medical society members saw no conflict between their rigorous standards and the democratic designs of the age. In fact, they championed their organizations as democratic. Admission was based solely on attainment, not hereditary right. Medical society members claimed to receive any man who managed his morals and learned the appropriate way to pursue medical science—any gentleman, in other words. Achievement of medical society admission criteria rested squarely on the individual practitioner. Nothing more than hard work and study were necessary to develop civilized habits and character.

The structure of early Cincinnati medical societies placed them in consonance with early nineteenth-century America. They emphasized the individual, stressing that each was a free agent and possessed both the ability and the opportunity to better himself. The medical societies of Cincinnati did not differ from medical associations elsewhere in early nineteenth-century America. All were structured similarly to Cincinnati's efforts, and all functioned in the same fashion and for the same purpose. These institutions appeared in concert with the notion of a fluid, democratic, individualistic America in which success appeared achievable and was measured in behavioral, not monetary, terms.[6] Attacks on these bodies as monopolies did nothing to lessen the fit. Those who cried privilege did not object to the idea, structure, or admission policies of medical societies. They complained only about the secrecy shrouding these organizations, an attack not unlike that on the Masons,[7] or about the unfair licensure law advantage that state legislatures sometimes conferred on these associations.

Early nineteenth-century licensure laws granted medical society members the exclusive right to choose new members, establish a societywide fee schedule, and sue for payment of fees, advantages no unlicensed practitioner could approach. Led by the Thomsonians—those individuals who had purchased the right to use the "medical discoveries" listed in Samuel Thomson's patent—and thundering "every man his own physician," antimonopolists objected to medical society licensing privileges in legislatures and courts. These crusades generally met with success, and by the 1830s most licensure laws had been overturned. Revocation of these trade restraints did little to alter medical organizations. Members mourned their loss of state sanction but returned to their now self-appointed duty of separating the medical wheat from the chaff. After all, sniffed Samuel Hanbury Smith, a member of the Royal Swedish College of Health and a powerful Cincinnati physician, "to be an M.D. according to law is one thing—to be a physician is another."[8]

It bears repeating that these medical societies took the individual physician as the locus of action, concern, and organization. The associations were viable precisely because they reflected their membership's and society's assumptions about democracy, opportunity, and success. Each individual practitioner could achieve civilization—become a gentleman—and attain society membership. Those who had demonstrated that they had developed the appropriate manners and temperament merited this achievement's perquisites. Those unable or unwilling to travel that road deserved to be left behind.

Beginning in the late 1830s, Cincinnati medical men, like their counterparts elsewhere, started to identify themselves in a new way. They stopped focusing on the merits of individual practitioners. In contrast, they now abruptly stressed instead the medical principles practiced, the system of medicine. Correspondingly, they began to consider themselves groups of practitioners distinguished by the medical system each group practiced. This new classification scheme changed relationships of individual practitioners to one another and to their discipline. The early nineteenth-century concern about which individual practitioners cultivated the most civilization—had made themselves the most civilized—was replaced by the question of which groups of physicians held the appropriate principles for American medical practice.[9]

This reidentification made itself felt immediately within medical organizations. Until the late 1830s, classifications of medical systems—regular, homeopathic, or eclectic—made little difference in the United States. Physicians espousing Samuel Hahnemann's homeopathic principles, emphasizing botanic remedies, or employing other varying *materia medica* (medical therapies) could be found peacefully coexisting in the same society. But after the shift from the individual practitioner to the group of principles used in medical practice, medical societies divided sharply along materia medica lines. Only a few of Cincinnati's most respected medical men—members of medical societies, Medical College of Ohio professors, or renowned practitioners—cast their lot with homeopaths or eclectics. Most chose to shun irregular systems of medicine and therefore avoided becoming irregular medical sect members; they opted instead to practice as regulars.[10]

Contemplation accompanied redefinition, and newly self-styled regulars found it easy to justify their choice. To those who characterized themselves as regulars, eclecticism and homeopathy seemed contrary to American values. The problem with these two most common irregular medical systems lay in their formulation. Both appeared grounded in theory, not observation-acquired facts. Homeopathy was based on the traditional European doctrine of *similia similibus curantus*—like cures like—and relied on infinitesimal dosing, whereas eclecticism stressed botanic remedies almost to the exclusion of all others. Both seemed rooted in hasty or faulty learning and erroneous reasoning; the two systems had arisen from an incorrect methodology and were therefore mere quackery. John Bell, professor of theory and practice at the Medical College of Ohio in 1850, put the regulars' case succinctly. "Like the poet who wrote a descriptive poem first, and then visited the spot which he describes," Bell charged that irregulars "first manufacture systems of medicine . . . and ethics . . . and then set about collecting facts and reasons for the support and elucidation of their vagaries."[11]

Regulars also condemned the conduct of eclectics and homeopaths, arguing that the same flaws that led irregulars to support an unsound medical system also resulted in tarnished personal behavior. Regulars chided irregulars for indulging in "flatteries, misrepresentations, and self-puffings" and "ostentatiously parading the number and respectability of their patients." They also accused irregulars of lacking the virtue of benevolence. Rather than work for social betterment, charged an irate regular, the quacks simply "fatten on the calamities and misfortunes of their fellow citizens" and limit "their efforts on behalf of humanity . . . to lining their own pockets."[12]

Regulars had a different notion of themselves. They claimed that they had cultivated mental discipline, benevolence, and modesty and that their medical system possessed a rich, progressive, and ancient history, which was nearing its apex in mid-nineteenth-century America. They argued that regular medicine began with Hippocrates, who laid down both the basic scientific tenets and the admirable code of conduct, and had been passed from notable doctor to notable doctor "in uninterrupted succession . . . down to the present generation." Cincinnati's regulars singled out Galen, Sydenham, Rush, and the archetypical American and much venerated Drake as bearers of a tradition that united "antiquity and universality," the foundation of which was "laid more than two thousand years ago."[13]

Redefinition of physicians in terms of groups set sects apart and in competition. This reidentification also possessed profound implications for each group's members. As sect members, physicians now found themselves benefiting from the good works or ominously blamed for the sins of all who claimed adherence to their medical system. The irregulars' attacks on regular medical therapies, especially purging and bleeding, assumed that regular physicians constituted a behaviorally homogeneous group, which translated into specific medical practices, and that each member was defined by those practices. Joseph R. Buchanan, professor of physiology at the Eclectic Medical Institute of Cincinnati, showed with devastating force that individual regulars received definition as sect members. He claimed that regulars purged and bled their patients maniacally and were little better than murderers. We eclectics recognize, Buchanan solemnly intoned, the "deplorable consequences [that] have been produced by the improper use of the materia medica, by the administration of harsh poisons, or pathogenic articles, in place of congenial, safe and restorative medicines." The regular physicians' "use of mercury and . . . calomel are gigantic evils," and "we war against them with energy and zeal."[14]

Regulars responded to attacks of this stripe with loud, defensive protests. The stinging rebukes of regular medicine implied to an already uneasy public both that all regulars employed an unconscionably extreme therapy and that calomel and the lancet constituted the regular materia medica's limits. Employment of but two remedies suggested a slavish adherence to dogma or doctrine—a contemporary likened the regulars' dependence on bleeding and purging to the popishness of Rome—an anathema to mid-nineteenth-century America.[15]

Warfare among medical sects sometimes involved more than verbal acrimony. Government-supported medical facilities often served as battlegrounds. For example, irregulars tried in the late 1840s to seize from regular domination Cincinnati's Commercial Hospital and Lunatic Asylum, run by the township trustees. The hospital's charter stipulated that its medical department be staffed exclusively by faculty members of the Medical College of Ohio, then the city's only regular medical school. Irregulars appealed to the legislature, demanding that each sect be allowed equal access to the facility. Irregular control of the hospital would be the inevitable product, agreed eclectics, because when each medical system is "fairly tested before the public" the results will "show the falsehood of the pompous pretensions of the incumbents to superior skill."[16]

Regulars recoiled at the irregulars' plan and in 1849 asked the trustees to testify to the legislature what fine medical care their sect had provided. They also forwarded a petition of support bearing 1,500 signatures, a ploy to which irregulars responded by sending a document with 10,000 signatures favoring equal opportunity. The irregulars' measure passed in the House, but regular medicine mobilized its forces in the Senate and the bill fell a few votes short.[17]

The 1849 legislative defeat failed to quell the irregulars' desire to control the hospital, and for much of the next decade their fruitless offensive persisted.[18] Irregular physicians did capture another Cincinnati medical institution, its health board. Regulars initiated the health board battle, and the 1849 cholera epidemic was crucial. The regular-dominated board precipitated a crisis when it ordered all city medical men to report to Health Officer C. S. Muscroft each case of cholera attended and the final disease outcome.

A staunch regular, Muscroft questioned the validity of the irregulars' reports. He claimed they inflated the number of cases treated and the percentage of cures produced, and Muscroft, with the board's consent, adjusted their statistics in his daily mortality reports. That miffed the irregulars, who now chose to sidestep the health officer and send their returns directly to city newspapers, which placed those sects in technical violation of the board's reporting clause. The health board ordered Muscroft to initiate legal proceedings,[19] and the health officer chose to arrest two irregulars as an indication of the board's seriousness. He selected Joseph Pulte and Benjamin Ehrmann, the city's two most esteemed homeopathic physicians. The arrest order backfired on the board when the mayor refused to uphold Muscroft's charges. The board's attitude to irregulars as well as its highhanded tactics emerged as the critical public issue. Cries from several quarters demanded the board's resignation. Council was called upon to reorganize the board "to avoid all prejudice" since "large numbers of our best citizens . . . refuse to swear by Mercury" and "public injury is sustained [by] this exclusiveness in the organization of the Board of Health."[20]

This complaint struck a responsive chord in City Hall, and council established a special health board committee. It proposed to replace the standing board with one composed exclusively of nonaligned laymen. Aware that "the different sects of religion are not so irreconcilable in their opposition to one another as are the opposing schools of the healing art," the committee offered its recommendation because

each sect regards the theory and practice of each other sect as not only wrong and injurious, but absolutely absurd; and religiously believes that those who follow it are either grossly deluded or dishonest. The pride of professional dignity has also raised up an impassable barrier between them, which bars even professional intercourse. With such implicit confidence in the correctness of his own theory, and in his own learning and experience, and practice of his antagonists, there is no ground on which the doctor of one faith and practice can meet and act with the doctor of another faith and practice.[21]

The health board resigned before council could act, chided the community for lack of support, and pledged not to serve on future health boards. Council accepted the resignations and immediately appointed a new six-member board. Gaining positions on this new board were James Taylor, who as editor of the *Cincinnati Times* had orchestrated the attack on the old board, and Bellemy Storer, Pulte's and Ehrmann's attorney. These appointments did nothing to reduce controversy as the new body proved as distasteful to regulars as the old board had to irregulars.[22]

Muscroft set the tone for his regular brothers. He refused to work with the new board, and when it dismissed him for insubordination, Muscroft declined to turn over health office records or meet with his successor. Muscroft's aggressive actions set some to wondering whether "the old school physicians, who composed in part, the old Board, [would] under the censure implied in this change, now report" cholera cases. The new board helped regulars decide when it removed the venerated Drake from his post as resident physician to the temporary cholera hospital post and replaced him with an eclectic. Regulars now "fully understood that the Board of Health was inimical to the regular profession" and having "been outraged and insulted by the dismissal of the old board, felt no inclination to bow to the new." They withheld their cholera statistics.[23]

The cholera reporting fracas cost regulars health board influence for the next decade. Loss of administrative control was hardly the only implication of the shift from individual practitioner to medical sect member. The new sense of relationship among each group's membership made them sensitive to every other group member's achievements and failures. The new conception of regulars, homeopaths, and eclectics as medical community members made it possible to worry about each community's coherence and homogeneity. That stance caused each community to discover and identify as problems unique to that community conditions and situations that had always existed in U.S. medicine. The very definition of physicians as members of medical sects opened the way after about 1840 to concern about each sect's lack of harmony and unity, the slim curricula and short terms at medical schools, the faulty preliminary education and correspondingly low character of many students and practicing physicians, and the tendency to eschew patient observation and leap toward grand theory. These "new" problems merged with the concern of each medical system's practitioners about the public's lack of respect for and confidence in their medical system. That latter concern translated into potentially decreased revenues and led each sect's members after 1840 to scrutinize intensely their medical community's internal condition to find ways to improve its public reputation.[24]

None of the "new" problems confronting the various medical systems' practitioners seemed to hold consequences as dramatic as absence of behavioral homogeneity among their sect's members. A sect could reap the rewards of public sanction only when its individual members behaved in a fashion to bring the entire group credit. The need to forge themselves into a homogeneous entity was particularly strong among regulars. Their assertion of a glorious ancient tradition proved a double-edged sword. Unlike irregulars who maintained that they had emerged as a reaction to and replacement for the regular system, regulars claimed a history that suggested that physicians formerly had acted in a homogeneous manner, possessed the public's respect, and had been rewarded with prestige and authority. "Discovery" of regulars behaving in a manner unbecoming to the sect prompted other, anxious regulars to explain this "new" condition and search for its origins, processes that pointed to potential "remedies."

Some commentators blamed regular medicine's lack of homogeneity on infiltration into its ranks of lazy and unprincipled men. "These parasites gain public confidence," complained Leonidas M. Lawson, professor of clinical medicine at the Medical College of Ohio and editor of *Western Lancet,* "by their forced associations—by claiming an affinity to a high and honorable profession," and their vulgar acts reflected badly "on the whole profession, and in the public estimation, the regular and scientific physician suffers by the unjust association thus established." Smith echoed Lawson's fears. He fretted about "the quacks in the legitimate profession" and asserted that "these black sheep" were numerous, standing as "deliberate traitors among ourselves, traitors to their profession . . . [and] society."[25]

Others attacked the false regulars' greed and vanity. Benjamin Franklin Richardson, later professor of obstetrics, gynecology, and pediatrics at the Miami Medical College of Cincinnati, identified "a class of men who are pursuing the profession merely as a trade, with very little thought except for its gains, with no wish to leave it better than they found it. Such ignoble souls," concluded Richardson, "are never willing to sacrifice the least private benefit for the general good." J. A. Murphy, soon to become professor of materia medica at the Miami college and coeditor of the *Cincinnati Medical Observer,* agreed but viewed inflated pride as the culprit. He argued that these men "seek to enter the profession that their silly vanity and self-love may be gratified by being saluted as 'Doctor,' that they may gain a position in society, and enjoy . . . the life of a physician."[26]

Depiction of individual regulars failing to uphold the sect's creed served as one explanation for heterogeneity. Other contemporary analysts attributed problems to a class who shunned traditional advancement methods, cultivation of mental discipline, and moral management. John P. Harrison, vice president of the nascent American Medical Association and professor of materia medica at the Medical College of Ohio, called for "every young man in the profession, whose primary education has been neglected, to repair his deficiencies, by the sedulous addictedness of his powers to the acquisition of the elements of a good English scholarship." The "vulgarizing effect of deficient scholastic education has already been felt," he warned, and without action by young regulars "the medical profession cannot maintain its true position in society."[27]

Some appalled regulars tried to dissociate themselves from individual practitioners who improperly claimed regular medicine affiliation or failed to achieve an appropriate standard. These perturbed physicians fell back upon the past's solutions and about 1850 established for themselves the Medico-

Chirurgical Society and the Cincinnati Medical Society. The two organizations promised founders to reestablish the individual basis of medical determination. They were testament to the idea that each individual could manage his own affairs and insure success and well-being. Except for the requirement that members champion regular medicine, the new societies proved identical to the predecessors, a manner of organization that caused their demise.

Similarities between the new and earlier organizations included admissions procedures. Only members could introduce a name into candidacy. An admissions committee examined a candidate's credentials and record to determine his worthiness and prepared a report. The entire organization then considered the nominee. The Medico-Chirurgical Society required a simple majority for admission, but the Cincinnati Medical Society proved more restrictive. Three negative votes blackballed a candidate.

Both societies also followed the well-established tradition of elegant, genteel, and rigorous meetings. Sessions took place at members' homes and featured a sumptuous repast, including such delicacies as "oysters, boned turkey and a liquid—prepared no doubt for medicinal purposes." Displays of cordiality and camaraderie soon gave way to a member's paper or case presentation. The societies tolerated no interruption of the address. The presiding officer called the roll after the talk and allowed each member in turn the opportunity to speak on the subject by standing and addressing the chair. Neither organization had a code of ethics because "none but a gentleman should be a physician, and being such would require no rules for his gentlemanly conduct." Their only rule of conduct was to punish anyone who during the comments forgot "the characteristics of a gentleman" and "descended to personalities, low epithets or indecorous language." Offenders were silenced and forbidden to continue until they made "a proper apology."[28]

Both societies appealed to members because of similarities to past medical associations. That nostalgic attractiveness destined them to fail. Medical sectarianism had replaced individual practitioners. Although a few faithful kept alive the new yet anachronistic societies for nearly a dozen years—they met but sporadically and only a handful attended—the organizations could never provide a solution to the problem for which they had been established. Unlike their chaff-distinguishing forerunners, which distinguished medical gentlemen from their less well developed contemporaries, the new societies operated strictly within the regular medical community. That placed them in the unenviable, even impossible, position of separating what had come to be seen by the public—the group that mattered—either as the wheat from the wheat or the chaff from the chaff.

Others attacked heterogeneity through modification of existing institutions, but sought transformation of defective community members, not exclusion—improvement, not separation. City medical schools stood as the initial site. Cincinnati's regulars formed in the early 1850s two new schools, the Cincinnati College of Medicine and Surgery and the Miami Medical College of Cincinnati. These institutions marked a fundamental medical school shift. The new medical education functioned as a means to upgrade regular medicine. Careful selection of students constituted a primary thrust. The three city medical school faculties discontinued the traditional practice of admitting any white male able to afford medical college tuition and restricted enrollments to students with solid belles lettres credentials and who had worked honorably under conscientious preceptors. Both course work and graduation requirements reflected the new impetus. Faculties now accentuated systematic study of medical texts and lectures as well as observation at the bedside. Patterned or systematic study inculcated in students the habits necessary to raise regular medicine's public standing. To insure that students had the time required to hone this skill, colleges set term length at five months, up from the previous four, and demanded that students attend the scheduled course of lectures for two years before taking the M.D. degree.[29]

Regulars understood that the new medical education constituted only a partial solution to the problem they faced. The great majority of regulars would have completed training prior to these reforms or never attended medical college. To reach segments of regular medicine unaffected by the new medical education and to induce harmony within the sect, Cincinnati's medical men did as their brothers throughout the United States did and stressed medical organization. Formulas for these institutions bore little in common with extant structures. These new agencies sought to include all regulars, regardless of politics or past conduct, within the fold. They would regulate regular medicine, systematizing relationships among practitioners and raising the grade of those not up to snuff.

Richardson set the tone. "Elevation of any body of men must proceed from within itself," and he cited teachers as a formerly degraded group now well respected. "If we ask how these changes have been effected," he wrote, "we shall find it has been by the united action of the teachers themselves. They have held conventions and institutes, formed societies, increased their own qualifications, and insisted on increased qualifications from those seeking admission to the ranks." Teachers, Richardson continued, "have educated the public mind to estimate at their true value, the services rendered by them. It is plain," he concluded, "that we are in the position formerly occupied by the teachers, and it is fair to infer that the same means which benefited them will benefit us."

Richardson recognized that approbation came from the public and realized that regulars would gain public support only if sect members seemed virtuous and could demonstrate these virtues. Heterogeneity among regulars had to cease. "Our profession cannot be honorable and respectable unless its individual members are educated to such a degree, and act in such a manner as to command the respect of the public," Richardson contended. Each individual regular acting in an untoward manner hurt the sect. Renegade or untutored regulars needed to be reined in, and he knew of "no more powerful means of watching and controlling individual conduct than afforded by a local society." Individual regulars, Richardson argued, "are the cells in whose proper or improper action depends the health of the whole." When these men are "united into societies they become tissues of the body medical, which possess the power of regenerating those cells likely to decay, and the casting off of those no longer necessary or beneficial to the general economy." If a regular acted improperly, Richardson saw "no recourse without organization." Acting "faithfully and cordially for their mutual good," society members would "compel the obnoxious individual to join the organization under the code of ethics . . . or take his place in the ranks of quackery."[30]

Sentiments like those expressed by Richardson struck a resonant chord among many Cincinnati regulars. It remained for Robert R. McIllvaine, a former president of the Medico-Chirurgical Society of Cincinnati and the Society of American Physicians in France, to bring the matter beyond discussion. In early 1857, McIllvaine returned from a trip to Paris and New York City and suggested at a Medico-Chirurgical Society meeting that the society disband and reconstitute itself as an organization patterned after the Academy of Medicine of Paris. While "in all matters pertaining to medicine, the medical profession in this country is not consulted by the public," he thought that "in France, the case was different." Since all practitioners must be French Academy members, "all matters of medical policy were referred to it, and its decisions were decisive." McIllvaine realized that Americans would not grant a medical society legal power to regulate medicine, but he argued for a society that would incorporate the best features of the academy concept. This Cincinnati academy would admit "all members of the regular profession," would not be "the attache of any clique or party," and would operate "under a liberal constitution." Open to public scrutiny, the "tribunal" would adopt a stringent code of ethics and hear cases of "merit and pretense," insuring that "each will receive its appropriate reward." It would force all swearing allegiance to regular medicine to behave in a manner conducive to their colleagues' and sect's well-being.[31]

The Medico-Chirurgical Society adopted McIllvaine's plan, but not without dissent. A minority found many regulars unworthy of medical society affiliation and questioned the wisdom of permitting laymen to view often heated medical debates. They preferred to maintain the status quo and refused to endorse McIllvaine's proposal or engage in the new society, which on March 5, 1857 became the Academy of Medicine of Cincinnati.[32]

The academy's founders quickly thrashed out the form for the new umbrella organization. The American Medical Association's code of ethics was incorporated into the academy constitution. Each member pledged on admission to uphold that code or be branded a quack. Regularly scheduled academy meetings emerged as a medium in which to improve deficient regulars. Beginning in 1858, the academy obliged each member to select a date and to deliver an address of his own choosing. Attendees checked and if necessary corrected the presenter's methodology. For members, being required to display their work and thoughts before their oft critical colleagues meant that the closely scrutinized essay served as potent force toward individual improvement, an activity now viewed within a group consciousness.[33]

Founders of the Cincinnati Academy of Medicine strove to extend organizational influence to students and beyond the meeting hall. Medical students, like laypersons, could not participate in academy proceedings, but medical college professors nonetheless required students to attend sessions. Officials suspected that academy attendance would imbue students with an appreciation of medical organization and insure postgraduate participation. The academy secretary routinely forwarded the society's proceedings to city medical journals, which granted members and nonmembers alike access to the organization's transactions for leisurely contemplation, study, or reflection. Spreading the word to the lay public stood behind the academy's policy of opening its doors to all parties. By allowing "all persons interested in the progress and status of our profession [to] listen to our discussions and witness our deliberations and transactions," explained E. B. Stevens, publisher and coeditor of the *Cincinnati Medical Observer* and the *Cincinnati Lancet and Observer,* the commonweal would be served. "The public, oft deluded by charlatanry, [will] receive profitable lessons, wise and salutary suggestions concerning legitimate medicine, its ethics and catholicity" and would inevitably throw its business to regulars.[34]

The academy was quite explicit as regards areas in which it would issue its profitable lessons and wise and salutary suggestions. Support for municipal regulation of and investigation into public health became the chief vehicle by which the academy tried to woo public opinion. Accentuation of public health implied that regular medicine followed in the American tradition of benevolence, a reputed national behavioral characteristic, and possessed the virtues necessary for medical practice in the United States. An occasional physician and some laypersons had ridden the public health horse earlier, but the academy remained the first organized body in Cincinnati to claim the cause as its own. Reuben Mussey set the society's tone in his inaugural oration. He identified the academy's "principal objects" as "investigation and discussions of such subjects as vital statistics, public and private hygiene, adulterations of food, progress in medicine and surgery, the condition of the atmosphere in relation to epidemics, original observations of disease [and] the encouragement of medical scholarship."[35]

Academy members translated Mussey's sentiments into action. Nine of the first sixteen academy presentations dealt with public health concerns, and they covered such diverse subjects as the evils of tobacco, sunstroke prevention and treatment, strychnine whiskey as cause of delirium tremens, vaccine disease, and milk adulteration. All came under the academy's scrutiny as part of its effort to demonstrate the indispensability of regular medicine to American society. The organization also expressly investigated the various epidemics that periodically raged in the city and repeatedly recommended that city government take dramatic action to rid Cincinnati of these plagues.[36]

The academy encountered some rough going in its early years but by 1860 was well established. It had become the city's sole regular medical society and had nearly one hundred members. The academy was catholic enough to include all regulars. It came complete with manner-regulating and morals-improving mechanisms, which boosted regular medicine's public standing. Murphy was especially buoyed by the academy's accomplishments. He contended just six years after its formation that the medical organization had "pressed out the bad influences" and produced "a single philosophy of medicine" among Queen City regulars.[37] Murphy overstated the case perhaps, but the academy's apparent success in raising regular medicine in the public imagination led to a corresponding decrease in gloomy rhetoric. Created to induce homogeneity among regular medicine practitioners, a move necessitated by the shift in identification from individual practitioner to group, the academy seemed likely to fulfill its objectives. It promised to convert the city's regulars into a homogeneous community and to convince the public of that community's cause. The academy appeared well on its way to becoming in mid-nineteenth-century Cincinnati "the voice of American medicine."

In a larger sense, the academy's formation and elaboration reflected the civil society of which it was so vital a part. The age of the individual had yielded to the era of the group, a change in perception that created new relationships, new interactions, and new dependencies. This definitional re-examination converted long-standing situations and conditions into resolution-demanding problems and made it possible to conceive of institutions from different premises to function in new ways. The Academy of Medicine of Cincinnati was one of those new entities, an exemplar of late antebellum American civilization.

PART TWO

The Age of Hierarchy, 1870–1920

Introduction to Chapter 4

In her discussion of puerperal insanity, Nancy Theriot provides a rich and satisfying analysis of the interrelatedness of larger cultural notions and technical knowledge. As Marcus has given us a powerful portrait of the shift of U.S. culture from the age of individualism to the age of the group, so Theriot's discussion of this important yet baffling affliction demonstrates again the importance of group identities and designations after the 1840s. The group, of course, was an invention of democratic culture, whether one thinks of post-1830s revivalism, the political party, the new understandings of corporations, or even that all-pervasive "associationist principle" that so dominated American cultural and social behavior in the middle decades of the nineteenth century.

And what was this strange disease? Puerperal insanity—literally, hysterical behavior patterns of new mothers following childbirth—appeared to be common in nineteenth-century America to most physicians, even though they described it in different ways at different times. It was responsible, Theriot indicates, for at least ten percent of female admissions to insane asylums, an arresting statistic indeed. Yet by the dawning of the twentieth century, it had all but disappeared. Like any other disease that "disappears," its history raises interesting questions. It can be seen, as Theriot depicts it, as a socially constructed illness and disease in the mid- and later nineteenth century, constructed in different ways by the different interests or groups involved in the phenomenon as a whole. For the women so afflicted, the proper perspective was the constrained roles that women played in society, including those with their medical doctors and alienists. For the doctors the illness was a manifestation of a category of patients and categories of professional expertise. For the woman's family, the identity of the patient, physician, medical establishment, and family itself were crucial. Theriot insists that puerperal insanity may be regarded as a socially and culturally constructed disease; it reflected both gender restraints and professional battles accompanying medical specialization. She notes that discussions of the affliction shifted from a relatively soft attitudinal phenomenon to a hard, seemingly biologically caused behavioral malady after the 1870s and that after 1900 discussions of the disease evaporated from the medical literature. Also, twentieth-century medicine was less tolerant of such categories as puerperal insanity. Thus, Theriot concludes, it may be inferred that the disease was the cultural construct or product of certain cultural notions in a particular age.

Theriot's study reminds us of another important shift in nineteenth-century American cultural life, that which the literary historians have called the transit from romanticism to realism, from an age of the transcendant spirit, or *Geist,* to an era in which the material aspects of life dominated sensibilities and perspectives. Indeed, the different medical systems of Marcus's Cincinnati doctors and the literary historians' depictions of transcendentalism in American literature would appear to be parallel phenomena in the same period, and the so-called moral reform movements of the three decades following 1830 might be thought of in the same light, for their champions defined America as a society and culture populated by many distinct groups and called for ideological proscriptions for the uplift of human behavior through the mechanism of wholesale behavioral reorientation. And Theriot's emphasis on the sudden redefinition of puerperal insanity in the later nineteenth century as having a biological base is congruent with much of what we know about cultural life in that era, including the widespread acceptance, among educated Americans at least, of the new evolutionary theories of Lamarck, Spencer, and Darwin, and the even more important veneration of expertise and science. It was in the five decades following 1870 that the genteel tradition died in American cultural life and that Americans from many walks of life acted and spoke as if a person's social standing depended on skill, talent, or merit (or lack thereof), instead of character or reputation, as had been the case since the later eighteenth century. Above all, as Theriot's essay suggests (and as much other historical literature fortifies), what was so remarkable was the new veneration of expertise and the expert, as manifested in the rising reputation of the new professionals in general but also scientists and engineers and doctors in particular—those important manufacturers and purveyors of technical knowledge.

Diagnosing Unnatural Motherhood: Nineteenth-Century Physicians and "Puerperal Insanity"

4

On December 16, 1878, Elizabeth S., age twenty-seven, was admitted to the Dayton Asylum for the Insane. The cause was puerperal; the form was mania. About three weeks before admission she had given birth, and her insanity appeared a few hours after the child was born. When Elizabeth was admitted to the hospital she was very noisy and excited, clapping her hands and talking incessantly. She would sometimes tear her clothing and expose her person. She had a poor appetite, did not sleep at night, and was in poor physical condition. Her physician insisted that she take plenty of milk and beef-tea every day, gave her iron three times a day, and thirty-five grains of hydrate of chloral (a sedative) at bedtime. Under this treatment Elizabeth remained the same for almost two months, except that she rested at night. Near the end of February, she began to improve. She started to take an interest in things around her, was more neat in her dress; thought she ought to have something better to wear, and would help do the work. She continued to improve and was removed from the institution by her husband on June 19, 1879.[1]

The case of Elizabeth S. was one of hundreds reported by physicians in nineteenth-century medical journals. Elizabeth's was a case of puerperal mania, the most common type of puerperal insanity. Physicians also described two other forms of the disease that usually had melancholic symptoms: insanity of pregnancy and insanity of lactation. Although doctors described puerperal insanity in various ways and although medical opinion about the nature of the malady changed over the course of the century, most physicians agreed that it was a common ailment and that it was responsible for at least ten percent of female asylum admissions. Yet, by World War I the

75

disease had all but disappeared. Except for postpartum depression, the twentieth-century renaming of insanity of lactation, puerperal insanity was cured by the world wars.

Like other nineteenth-century female diseases that have disappeared or been redefined in the twentieth century, puerperal insanity raises many questions about the relationship between the predominantly male medical profession and women patients. Was puerperal insanity an invention of men? Was it an expression of male physicians' ideas about proper womanly behavior, defining women's antimaternal feelings and activities as insane? Or was puerperal insanity only incidentally a gender issue; could it be understood as a professional struggle between male gynecologists and male alienists (nineteenth-century psychiatrists) over the treatment of insane women? Given the sexual politics involved when women's illness is named and treated by a male medical establishment, can physicians' accounts of puerperal insanity provide valid information about the meaning of the disease for women? If so, was puerperal insanity an indication of dissatisfaction with motherhood, disappointment with marriage, or anguish over abandonment or financial problems? In short, was puerperal insanity an expression of sexual ideology, medical professionalization struggles, or gender tension?

These questions cannot be answered adequately using either the traditional approaches to the history of insanity or the more critical approaches taken by historians interested in the history of women and madness. Both traditionalists and critics explain nineteenth-century insanity (or specific insanities) from one of three perspectives: that of the disease, the physician or medical institution, or the patient. Each vantage point is important, but incomplete.

Although concentrating on the disease itself can provide information essential to interpretation, disease-focused studies deal with the disease either as an idea or as an essence gradually becoming known or named. Treating insanity or insanities as histories of ideas is interesting and useful, but this approach sidesteps questions of power.[2] Understanding how the idea of puerperal insanity changed over time and how it related to other insanities is essential, for example, but this understanding does not begin to answer the questions posed earlier about gender and power. Similarly, it would be a mistake to see puerperal insanity as a "real" disease, misunderstood or misnamed by nineteenth-century physicians, but understood and rightly differentiated by twentieth-century psychiatry.[3] This approach to insanity or insanities ends up begging all the questions of the meaning of insanity: why was this set of symptoms seen in a particular way at this particular time? Why was this group of patients seen as at risk? Why was this disease named one way in 1850 and another way in 1910? Interpreting changing insanities as a change in medical nomenclature leaves all of the important questions not only unanswered but also unasked.

Interpreting the history of insanity from the perspective of physicians or medical institutions is more fruitful than the disease-centered approach because focusing on the medical establishment demands that insanity be situated within a specific socioeconomic setting. From this point of view the reality of the disease is questioned or ignored, as the historian concentrates on the role of professional and institutional politics or individual physicians in the creation of insanity. Perhaps the best known example of this approach is *Madness and Civilization* (1965) in which Michel Foucault argues that medical discourse on insanity helped to define "reason" by medicalizing and silencing an ever-increasing category of "unreason."[4] Similarly, many twentieth-century medical sociologists see insanity as a label applied by a powerful medical establishment to society's deviants.[5] Historians writing about nineteenth-century insanity have also noted the role of professional rivalries between alienists and neurologists in defining the nature of insanity, as well as the role of individual physicians (such as George Beard and S. Weir Mitchell) in discovering, classifying, and treating insanities.[6] What all of these approaches share is an emphasis on the power of organized medicine to define certain behavior as insane or neurotic.

Many feminist historians and sociologists writing about women's insanity have concentrated on the power of physicians to categorize women's behavior as normal, neurotic, or insane, and have pointed out how such categorizations both reflect and help maintain gender stereotypes and the imbalance of power between women and men.[7] While this perspective is superior to a disease-focused approach because it makes visible the sexual politics of medicine, there are problems with the physician-oriented interpretation. A major difficulty with concentrating on the medical establishment as the creator of insanity categories, or as the agent of society in its quest to control deviants, is that patients, the public, and/or women are seen as passive victims of medical definition. Reducing insanity to a behavior pattern defined as sick by a powerful profession tells us little about the meaning of that behavior in the lives of the patients.

Since Carroll Smith-Rosenberg's early article on hysteria, some feminist historians have interpreted women's insanity from the point of view of the patient, asking what the symptoms meant to the women afflicted. Like the physician-oriented perspective, concentrating on the meaning of the disease for the patient involves situating insanity in a particular cultural location. Smith-Rosenberg's study, and a later study of anorexia by Joan Jacobs Brumberg, interpret the illness within a specific family dynamic: woman as wife or daughter in a constricted or contradictory life pattern.[8] This patient orientation moves away from the essence of the disease and the politics of defining it and instead asks why a woman might have behaved in a certain way. When trying to understand women's insanity it is absolutely essential to focus on the meaning of the behavior within the context of women's lives, but there are at least two risks involved in relying solely on this perspective. "Insane behavior" might be misconstrued as heroic, as the only sane thing to do when confronted with a particular life situation. And, in concentrating on the family dynamics or the specific gender constraints of the patient, one might miss the medical dynamic and the process of defining and labeling behavior as insane or abnormal.

In order to understand the relationship between gender and insanity in general, and puerperal insanity in particular, we need a method of analysis that can encompass all three perspectives—those of the disease, the physician/medical establishment, and the patient—and that describes the three in dynamic interrelationship. We need an interpretation that will be able to offer an explanation of both the meaning of symptoms in the lives of patients and the translation of symptoms into disease categories by medical professionals. What follows is an interpretation of puerperal insanity that divides the symptoms into illness and disease and sees both as social constructions.[9] The illness of puerperal insanity was a behavior pattern expressing dissatisfaction or even despair over the constraints of womanhood in a particular time; while the disease of puerperal insanity was a definition given by physicians to the illness symptoms, a definition that both legitimized the behavior pattern and played a role in medical specialization. As both illness and disease, puerperal insanity involved relationships: between the woman and her family, between the woman and her doctor, between the husband and the doctor, and between different medical specialists. Puerperal insanity can be interpreted as a socially constructed disease, reflecting both the gender constraints of the nineteenth century and the professional battles accompanying medical specialization. Male physicians and their female patients, together, created puerperal insanity; and that creation both reflected and contributed to sexual ideology and medical specialization.

Before elaborating this interpretation, a more thorough examination of puerperal insanity is in order. As mentioned earlier, most physicians believed puerperal insanity manifested itself differently in the three phases of the reproductive process. Milton Hardy, the medical superintendent of the Utah State Insanity Asylum, defined puerperal insanity as a condition developing "during the time of and by the critical functions of gestation, parturition, or lactation, assuming maniacal or melancholic types in general" and characterized by "a rapid sequence of psychic and somatic symptoms which are characteristic not individually, but in their collective groupings."[10] Some physicians preferred to classify puerperal insanity as maniacal, melancholic, or depressive, instead of dividing it according to reproductive phase; but in both groups there was consensus as to the type of insanity most associated with pregnancy, parturition, and lactation.

Insanity of pregnancy was thought to be the rarest of the three, and usually involved melancholic (and suicidal) symptoms or depressive symptoms. Nineteenth-century physicians described as melancholic patients who appeared to be apathetic, hopeless, and prone to suicide, whereas depressive patients were those with low spirits. In cases of insanity of pregnancy, the symptoms sometimes lasted only a few weeks or months, but in other cases the patient was cured only by childbirth. Insanity of pregnancy was thought to occur most often with first pregnancies; however, some women who had developed symptoms once would develop symptoms in subsequent pregnancies. This form of puerperal insanity was rarely fatal.[11]

Lactation insanity was similar to gestation insanity in its symptoms, melancholic and depressive, but was seen as more frequent. Lactation insanity differed from insanity of gestation and parturition in that it seemed to occur most often in women who had several children rather than in women going through their first pregnancies. In some cases of lactation insanity, the melancholy ended in dementia and lifelong commitment to an asylum, but most cases recovered in under six months.[12]

Insanity of parturition was considered the most common type of puerperal insanity and was associated with maniacal symptoms. Usually puerperal mania began within fourteen days of childbirth, but some cases started up to six weeks later. Like the insanity of pregnancy and lactation, puerperal mania was rarely fatal and usually lasted only a few months. Of the three forms of puerperal insanity, puerperal mania was the most baffling to medical writers in the nineteenth century. Indeed, most of the medical literature on puerperal insanity was a description of puerperal mania. Characteristic symptoms included incessant talking, sometimes coherent and sometimes not; an abnormal state of excitement, so that the patient would not sit or lie quietly; inability to sleep, with some patients having little or no sleep for weeks; refusal of food or medicine, so that many patients were fed by force; aversion to the child, the husband, or both, sometimes expressed in homicidal attempts; a general meanness toward caretakers; and obscenity in language and sometimes behavior.[13]

Until the end of the century when doctors began to express suspicion about puerperal insanity as a specific illness, there was widespread agreement about its frequency, duration, and prognosis. A physician writing in 1875 asserted that puerperal insanity was "a class of cases to be met with in the practice of nearly every physician," others cited asylum records indicating that the disease was responsible for "a very large proportion of the female admissions to hospitals," and still others claimed that puerperal insanity affected from one in four hundred up to one in one thousand pregnant women.[14] Physicians from every region of the country reported cases of puerperal insanity and noted that the disease struck middle-class women in comfortable homes as well as unwed mothers living in poverty. Doctors also agreed that most cases of puerperal insanity lasted only a few months, with most patients recovering completely within six months.[15] Except for those cases with suicidal or homicidal tendencies, the prognosis was good for patients suffering from puerperal insanity, and doctors asserted that most cases could be, and were, treated at home.[16]

Treatment for puerperal insanity remained mostly the same over the course of the century, and the change reflected a more general change in medical therapeutics. In the first part of the century, bleeding was considered the proper treatment, no matter if the symptoms were manic or melancholic. By mid-century, that treatment was no longer recommended, and instead physicians were treating puerperal insanity patients with rest, food, a little purging, and sedation. Most physicians also recommended that patients be restrained or watched closely and that family and friends be kept away.[17]

One of the first explanations of puerperal insanity to occur to an historian sensitive to gender as a category of analysis was that the disease represented male physicians' definitions of proper womanly behavior.[18] To nineteenth-century men, a woman who rejected her child, neglected her household duties, expressed no care for her personal appearance, and frequently spoke in obscenities had to be "insane." Certainly there is much in the medical literature to support this explanation. Many physicians wrote in very sentimental terms about the mysterious beauty of motherhood being defiled by insanity. Dr. R. M. Wigginton wrote of the special horror of puerperal insanity: "The loving and affectionate mother, who has so recently had charge of her household, has suddenly been deprived of her reason; and instead of being able to throw around her family that halo of former love, she is now a violent maniac, and feared by all."[19] Physicians commented on a woman's "letting herself go" or being "indifferent to cleanliness" as symptoms, and many listed willingness and ability to perform household tasks as evidence of a cure.[20]

By far the most shocking symptoms of puerperal insanity were women's indifference or hostility to children and/or husbands and women's tendency to obscene expressions. The first upset physicians' ideas about women's maternal and wifely devotion, while the second undermined doctors' assumptions about feminine purity. Allan McLane Hamilton described a patient who before her labor was "a loving and devoted wife, but shortly after lost all of her amiability, and treated her husband and mother with marked coldness, and sometimes with decided rudeness."[21] Even more difficult to explain than coldness was a woman's "thrusting the baby from the bed, disclaiming it altogether, striking her husband," a woman who looks at her baby "and then turns away," or a woman who "commenced to abuse it [the newborn child] by pinching it, sticking in pins, etc." So frequent was hostility or aversion to husband and child noted in cases of puerperal insanity that this was considered one of the defining characteristics of the disease, and physicians recommended that the woman not be left alone with her infant.[22]

If doctors were horrified at women's treatment of husbands and children, they were equally shocked at women's obscene words and behavior during an attack of puerperal insanity. "The astonishing familiarity of refined women with words and objects and practices of obscene and filthy character, displayed in the ravings of puerperal mania, gives a fearful suggestion of impressions which must have been made upon their minds at some period of life," wrote George Byrd Harrison, a Washington, D.C. physician. W. D. Haines of Cincinnati described a case in which the woman would repeat one word a dozen or so times "then break forth into a continuous flow of profanity. The subject of venery was discussed by her in a manner that astounded her friends and disgusted the attendants." Another doctor wrote of the typical puerperal mania patient "tearing her clothes, swearing, or pouring out a stream of obscenity so foul that you wonder how in her heart of hearts such phrases ever found lodgment." An Atlanta physician expressed similar puzzlement: "It is odd that women who have been delicately brought up, and chastely educated, should have such rubbish in their minds." And still another physician described this symptom as "a disposition to mingle obscene words with broken sentences . . . modest women use words which in health are never permitted to issue from their lips, but in puerperal insanity this is so common an occurrence, and is done in so gross a manner, that it is very characteristic." W. G. Stearns, a Chicago physician, went so far as to note that in "all such cases [puerperal mania] there is a tendency to obscenity of language, indecent exposure, and lascivious conduct."[23]

Clearly, these physicians were shocked and dismayed by their patients' "indecent" behavior and use of language, as well as by their hostility toward husbands and infants, their neglect of household duties, and their refusal to pay attention to personal appearance. Even in their empirical reporting of patients' symptoms, doctors revealed their disgust and horror over such unwomanly women. In naming their behavior puerperal insanity, physicians were both reflecting and supporting nineteenth-century sexual ideology.

As authoritative spokesmen for the new scientific view of the nature of humanity, physicians were also helping to create sexual ideology in their explanations of puerperal insanity. Many doctors wrote of insanity as a logical by-product of women's reproductive function. George Rohe, a Maryland physician, asserted that "women are especially subject to mental disturbances dependent upon their sexual nature at three different epochs of life: the period of puberty when the menstrual function is established, the childbearing period, and the menopause."[24] Dr. Rohe regarded insanity as an ever present danger to all women throughout their adult lives. Other doctors, however, wrote of pregnancy as a special challenge to women's mental balance, asserting that most women suffer mild forms of mental illness throughout their pregnancies. "In females of nervous temperament, the equilibrium of nerve force existing between these two organs [the brain and the uterus] is of the most delicate nature," wrote a Denver physician. He went on to say that "pregnancy is sufficient to produce insanity."[25] Probably the clearest statement along these lines was made by a professor of gynecology who wrote: "From the very inception of impregnation to the completion of gestation, some women are always insane, who are otherwise perfectly sane." He went on to say that others manifest defective mental integrity in the form of whimsical longings for the gratification of a supposed depraved appetite.[26]

It would seem that nineteenth-century physicians' views of proper womanly behavior, along with their ideas about the power of the uterus to disrupt women's mental balance, influenced their perception and definition of puerperal insanity. It would be a mistake, however, to conclude that puerperal insanity was simply an indication that male doctors reflected their time or that the medical establishment influenced sexual ideology. Focusing too closely on the obvious ideological content of physicians' accounts of puerperal insanity, one might overlook that physicians' guesses about the nature of the disease were very much in keeping with nineteenth-century ideas about insanity in general and that many physicians offered what late twentieth-century people would call sociological explanations for women's behavior. Indeed, much of the medical discourse on puerperal insanity seems to have been influenced very little by male doctors' concepts of femininity, but instead reflected the state of medical knowledge about insanity, on the one hand, and a jurisdictional dispute between alienists and gynecologists over the treatment of insane women, on the other.

For example, throughout the nineteenth century physicians asserted that mental illness in general, not just women's mental illness, reflected a connection between mind and body; if the mind was unbalanced, a brain lesion was responsible, and the exciting cause of the brain lesion could be physical or emotional.[27] Indeed, this argument was one of the ways physicians convinced the public that mental illness was a medical problem. From the general assumption of a mind-body link as part of the nature of mental disease, it was logical to conclude that puerperal insanity was in some way caused by the physical state of pregnancy, parturition, or lactation. Doctors reasoned that the physical system was taxed by the reproductive process and that this added strain could be an exciting cause of insanity. A Pennsylvania physician wrote that "[t]here is no organ or portion of viscera which is not intimately connected with the brain through the sympathetic nervous system," and the Ohio physician who admitted Elizabeth S. to the Dayton Asylum noted more specifically about puerperal insanity: "The physical derangements attendant upon pregnancy, child-bearing and nursing, are the principal causes of the insanity, which would be equally produced by any other physical suffering or constitutional disturbance of the same intensity."[28] Another indication of this line of reasoning was physicians' notation of any physical problem associated with labor as the probable cause of the insanity. If there was infection or a mild fever, if the labor was unusually long or difficult, if the woman required forceps, if her perineum was torn: these were seen as explanations for the puerperal insanity.[29]

Physicians also cited "heredity" as a primary cause of puerperal insanity, especially by the middle of the century. Like the mind-body theme, this too reflected a more general trend in medical ideas about the nature of mental illness. If there was insanity in a woman's family, regardless of how remote a relationship, this was considered a predisposition to mental unbalance. In such a case, pregnancy, childbirth, or lactation was seen as the stress that pushed the already unstable mind over the edge.[30]

Finally, many physicians argued that puerperal insanity was caused by situation, what the nineteenth-century writer called moral factors and what the late twentieth-century writer would call sociological factors. This too was in keeping with nineteenth-century theories about insanity in general. Just as financial problems or job stress were seen as possible causes of insanity in men, women were thought to develop puerperal insanity sometimes because of being abandoned or poorly treated by husbands, being pregnant and unmarried, being overburdened with too many children and household cares, or being emotionally drained because of grief or fear. In such cases physicians were very clear that the woman's insanity was brought on by her situation and that the puerperal state simply lowered the woman's strength so that she could no longer deal with the adverse environmental conditions. Kindness, rest, and reassurance were the best treatment.[31]

The mind-body connection and the possibility that physical or moral factors could be the exciting cause of puerperal insanity were both stressed throughout the century, but by the 1870s gynecologists began to emphasize the physical causes. The earliest proponent of this point of view, cited later as a man ahead of his time, was Horatio Storer. He argued as early as 1864 that most insanity in women is reflex insanity; that is, the primary cause of the insanity is a malfunction of the reproductive organs. For Storer and his post–Civil War followers, this meant that women's insanity could be prevented, treated, and cured by medical and/or surgical means.[32] It also meant that a gynecologist should be consulted in any case of female insanity. Medical ideas about the nature, cause, and treatment of puerperal insanity were complicated by this professional struggle. Because it was in their best interest to link women's insanity with their reproductive organs, gynecologists saw a connection that other physicians saw less clearly. Furthermore, they wrote authoritatively, as the "medical experts" on women, and assumed disagreement was the result of ignorance. Charles Reed, professor at the Cincinnati College of Medicine, expressed surprise to hear any dissent from the "long-recognized doctrine of the genital origin of insanity in the female sex."[33] One Washington, D.C. physician claimed that puerperal insanity could be prevented only by good prenatal care.[34] These gynecologists directed their arguments to general practitioners and to alienists, who ran asylums. Many, though not all, of the gynecologists' articles about puerperal insanity or about women's insanity in general concluded that asylums should employ gynecologists—a clear expression of the professional struggle influencing medical perceptions of women's insanity.[35]

The medical discourse among gynecologists, alienists, and general practitioners about the nature of female insanity affected practice, which in turn affected discourse. From the mid-1870s to the 1890s gynecologists practiced their medical and, increasingly, surgical techniques on private patients and institutionalized women. Increasingly diseases of the reproductive system were listed as the cause for the insane symptoms of women admitted to asylums.[36] More and more asylums employed gynecologists to examine female patients upon admission, and physicians found a variety of gynecological disorders among the women. Believing that there was a direct connection between these disorders and the women's insanity, the doctors administered medical and surgical cures. In the surgical category, removal of the ovaries was the most popular operation, but more and less extreme operative procedures were also tried, such as hysterectomy and birth repair surgery.[37]

Some physicians reported patients being cured of insanity as a result of a gynecological procedure, and puerperal insanity was said to be especially responsive to physically oriented therapy. However, as gynecologists treated more insane women, in and out of asylums, medical discourse reflected their growing disillusionment with surgical and medical treatment. Even those physicians who supported operative treatment reported disappointing cure rates.[38] By the 1890s there was lively debate over surgical treatment of insane women, with some physicians denouncing "mischievous operative interference" and others asserting that only physical (not mental) symptoms should prompt a surgical response. What made the debate different from the earlier one in which gynecologists successfully fought for the right to treat insane women was that the later debate was based on empirical studies. Having won access to asylum patients, gynecologists generated the numerical evidence against their own case. Two Minnesota physicians working at the state hospital at St. Peter found a large number of women asylum patients with serious pelvic disease in whom "there was not only no apparent relation between the pelvic disease and the mental disturbance, but there was no complaint or evidence of physical discomfort." They called this finding "the most unexpected result of our investigation."[39] Other physicians recorded the effect of surgery on women's insanity and found no significant link between operations and cures. Although they argued that gynecological problems could add to a woman's worry and discomfort and that all women (in and out of asylums) should have those problems treated, most gynecologists by the end of the century no longer claimed that women's diseased reproductive organs caused their insanity.[40]

If the empirical evidence, most of it gathered by gynecologists themselves, would not support a straight physiological explanation of women's insanity, how were physicians to account for puerperal insanity? Gradually, beginning in the 1890s, puerperal insanity was seen as a suspect category, and the emerging specialty of psychiatry emphasized the similarity between puerperal mania and any other mania, between the melancholy some women experienced during pregnancy or lactation and any other melancholy.[41] The particular physiological process was seen as less and less significant, and so the very term *puerperal insanity* was eventually dropped. Just as its appearance and growth was complicated by struggles of medical specialization, the disappearance of puerperal insanity from medical discourse was due to the empirical studies of one specialty and the reconceptualization of insanity that accompanied the rise of a new specialty (psychiatry).

Seen from this angle, puerperal insanity was not simply an expression or creation of sexual ideology by the medical profession. Certainly gynecologists were able to convince other physicians of the physiological basis of women's insanity (and puerperal insanity) because the argument fitted common ideas about woman's nature. Physicians saw mad women in a particular way because of generally held cultural ideas. That medical discourse was altered by empirical investigation at a time when most Americans, including feminists, believed in a biologically determined woman's nature indicates that gender was not a simple factor in the medical debate. Perhaps the most significant way gender affected the medical construction of puerperal insanity is in the absence of women from the professional discourse until the late nineteenth century. There is no way to measure the impact of women's silence, but it is interesting to note that women physicians in the 1880s and 1890s were overrepresented in the group of doctors gathering evidence that separated women's insanity from their reproductive organs and eroded the assumptive framework for puerperal insanity as a specific illness.[42] It is safe to assume that the exclusion of women from medicine in the early and mid-nineteenth century affected the scientific view of women's mental (and physical) illness.

So far we have been concentrating on physicians and the ideological and professional issues influencing their conception of puerperal insanity. The medical discourse, however, also offers a way to understand the women who were patients. Most medical articles dealing with puerperal insanity included case studies, detailed descriptions of the situation, behavior, and treatment of the patients. Of course, what doctors selected as important information and what they recorded and did not record of patients' speech and behavior was subjective. Yet they were attempting objective observation. Although we cannot take case studies as the complete picture or as entirely unbiased accounts, they reveal much about the possible meaning of puerperal insanity to the women who were so diagnosed, and they also provide a somewhat blurry snapshot of the doctor-patient dynamic.[43]

On the most literal and superficial level, case studies of puerperal insanity indicate that many women responded with melancholic or maniacal behavior to situations that they found unbearable. Illegitimacy, the fear that often accompanied first pregnancies, a traumatic birth experience or a stillborn infant, infection following birth, and extreme cruelty of husbands—were all cited in case studies, sometimes with the doctor attributing the insanity to the situation and other times not. One woman developed maniacal symptoms after her baby was delivered with forceps ("the head was extracted with considerable difficulty"), and she suffered physical damage in this, her first delivery. Another woman "frail and feeble" developed insane symptoms after her infant died a few days after birth. A woman whose symptoms included disclaiming her infant, striking her husband if he came near, and accusing people of trying to kill her was unimproved after five months in an asylum; her baby had died two months earlier, and her husband, it turned out, had been continually abusive to her during her pregnancy.[44]

Other situational difficulties also appeared in case studies, such as women having many children in very few years and seeming overburdened with work and responsibility. One woman, Mrs. S., who was thirty-five and had had five children, three of them within five years, developed "anxiety and slight confusion of ideas" during her last pregnancy. After the child was born she went into a "furious delirium . . . tried to leap from the window to avoid imaginary pursuers." A few days later she was no better; "she said she expects to be tortured soon, remonstrates bitterly." By the tenth day she was a little better: "Talks less and sleeps better. Tries to explain her sickness but cannot."[45] In another case a twenty-two-year-old woman was melancholic after the birth of her fourth child; her husband confined her and abandoned her once she was hospitalized.[46]

Case studies of puerperal insanity almost always included some physical or situational problem that late twentieth-century readers would see as cause enough for insane behavior, even when the physicians failed to note the connection. But while we may conclude that these women had good reason to act strangely for a few months, the meaning of puerperal insanity is more complicated than this. The symptoms provide a clue to the meaning of the disease for women and also point to the doctor-patient relationship as a key factor in the waxing and waning of puerperal insanity.

Whether on a conscious or unconscious level, women who suffered from puerperal insanity were rebelling against the constraints of gender. The symptoms clearly indicate that rebellion. Case studies document that women refused to act in a maternal fashion by denying their infant nourishment or actively attempting to harm the child. Many women "did not recognize" the child, "ignored" its presence, or denied that the child belonged to them.[47] Similarly, women refused to act in a wifely fashion; they claimed not to know their husbands, expressed fear that the husband wanted to murder them, and sometimes struck out physically at their husbands.[48] Women were refusing the role of wife and mother, a role that most nineteenth-century Americans saw as the essence of true womanhood.

Moreover, women suffering from puerperal insanity were not acting like women at all. They were apathetic, irritable, gloomy, and violent, instead of tuned in to the needs of those around them. In fact, these women required that others pay attention to them, in their constant talking and pacing the floor and in their refusal to care for themselves in the simplest ways, such as feeding themselves and keeping themselves clean. In a time when modesty was thought to be a defining characteristic of femininity, women with puerperal insanity laughed immodestly, tore their clothing in the company of men, and used obscene language. Rebellion against cultural notions of true womanhood was the one thing tying together the various symptoms of puerperal insanity.

Physicians, new to the lying-in chamber, made these rebellious symptoms legitimate by defining them within a medical framework. Doctors responded to women's behavior with a name: puerperal insanity. That naming was the result not only of the general ideas of the culture and the specific professionalization struggles of physicians, but also was related to doctors' new relationship with women patients: as birth attendants. From the late eighteenth century, male physicians had begun to describe pregnancy and childbirth as a traumatic ordeal. Even doctors who did not think of birth as a sickness but described it as a natural phenomenon expressed a mixture of amazement, disgust, and respect at women's ability to undergo all the physiological changes associated with pregnancy, birth, and lactation. The assumptions of nineteenth-century physicians provided a framework for both their acceptance of women's strange behavior as a side effect of reproduction and their definition of that behavior as, mostly temporary, insanity.

The medicalization of pregnancy, birth, and lactation provided a kind of permission for women to express rebellion and desperation in the particular symptoms of puerperal insanity. But if physicians and women patients both participated in the creation of puerperal insanity, the relationship was not a straightforward one. Women played out their rebellion against the male physician, and doctors translated that rebellion into an acceptable medical category. But doctors also cured the rebellion with their treatment and systematically silenced women in their case study reporting. In both cases, women were unequal partners in the construction of the disease.

Treatment of puerperal insanity consisted of various levels of constraint and intrusion. In what late twentieth-century readers would judge the mildest, most humane treatment, women were confined to their rooms, denied the company of family and friends, and forced to rest by the admission of tranquilizers. If the woman refused to eat, which happened in an overwhelming majority of puerperal insanity cases, she was force-fed. Indeed, the element of force was characteristic of most treatment plans. One physician recorded force-feeding and threatening to cut the patient's hair if she continued to refuse food, and others noted that patients were confined to their rooms or their beds if their behavior did not change quickly enough.[49] In nonsurgical cures force-feeding was the most intrusive aspect of the treatment, but surgical cures were penetrating in a more drastic sense. For the doctor, these cures were restoring the unfortunate patient to her rightful and happy role. For the woman? Regardless of how women perceived the cures, and we will never know their perceptions, they certainly gave up their insane behavior, usually within a few months. If women were expressing rebellion in puerperal insanity symptoms, and male physicians were defining that behavior as medically explainable and therefore legitimate, male physicians were also forcefully putting down the rebellion. In the social construction of puerperal insanity both parties were not equally powerful. A more interesting example of women's subordinate position in the relationship defining puerperal insanity is the judging and editing of women patients in the male-controlled medical discourse. The language physicians used to describe their women patients was often sympathetic but more often judgmental. One doctor described a woman before her insanity as having a "naturally obstinate and passionate disposition," and another wrote of a suicidal mother who tried to harm her four-month-old infant: "She should be hung." [50] More subtle than judgments of behavior were descriptions in case studies that substituted judgment for information. Physicians recorded "obstinate" and "indelicate" behavior and "immoderate" laughter. In some cases the physicians' judgmental words were simply reflections of husbands' accounts of their wives' behavior; but that acceptance of the husbands' point of view was very much a part of the sexual politics involved in puerperal insanity. To many male physicians, the women were to blame for their deviant, unwomanly behavior, and physician case studies recorded the blame.

Although women patients and male physicians constructed puerperal insanity together, the clearest indication that men controlled the discourse was the near absence of women's words from the case studies. Over and over again physicians claimed that women suffering from puerperal insanity "talked incessantly" yet no attempt was made to record what the women talked about. Similarly, some women were said to complain of "imaginary wrongs," with no explanation of the content of those complaints. The most glaring omission in the case studies was physicians' refusal to record women's "obscene" language. An overwhelming majority of case studies referred to one or all of these speech acts, yet no content was provided.

If women were silenced partners in the construction of puerperal insanity, what can we conclude about the meaning of the disease for women? Although women's words were not reported, physicians' accounts of women's behavior and situations indicate that puerperal insanity was an unconscious act of rebellion against gender constraints for many women. The particular symptoms of puerperal insanity involved a denial of motherhood and a reversal of many feminine traits. Women presented these symptoms and acted out their rebellion; male physicians who for ideological and professional reasons were disposed to define women's behavior as insanity, legitimized women's rebellion as illness. Yet part of the meaning of puerperal insanity for women must also have been the curing, the silencing. So many of the symptoms were aggressively, willfully expressive: the tireless pacing, the continuous talking, the laughter, the obscenity—all unlistened-to, unrecorded. It is almost as if women usurped the power of language only to find that it held no power at all. The woman cured of puerperal insanity surrendered these self-assertive symptoms and went back to being the halo of love in her family, without having been heard. There is no way of knowing whether she saw herself as victorious or defeated.

In spite of the sexual politics inherent in the doctor-patient relationship defining puerperal insanity and in spite of women's silence in the case studies, women's symptoms were taken seriously enough to constitute a disease, at least until the turn of the twentieth century. What did it mean for women that puerperal insanity disappeared? Certainly it can be argued that the constraints of gender were not as tight in the early twentieth as they had been in the nineteenth century. Women were having fewer children, childbirth was less dangerous and less painful, women had wider opportunities in terms of education and work, and women's marriage relationships were more compassionate. If puerperal insanity was a rebellion against the constraints of nineteenth-century "true womanhood," women may have had less trouble with the twentieth-century variety and therefore ceased to manifest the symptoms of puerperal insanity.

Although changes in women's situation contributed to the demise of puerperal insanity, changes in the relationship between doctors and women patients also played a part. As we saw earlier, empirical studies and the rise of psychiatry altered medical perception of mental illness. Reliance on more objective, scientific studies as the basis of medical discourse meant that there was less tolerance for puerperal insanity as a category. Regardless of how much or little women's situation had changed by the twentieth century, the symptoms of puerperal insanity were no longer a legitimate response to pregnancy, birth, or lactation in 1910, as they had been in 1870. Changing medical ideas, which had little to do with women patients, meant that physicians would no longer legitimize puerperal insanity as illness.

Elizabeth S. was admitted to the Dayton Asylum For The Insane in 1878. Her illness was the product of several intertwined relationships: her own response to her marriage and motherhood; her physician's response to her story; and her story's resonance in the medical and cultural score of the nineteenth century. The interaction of these layers of relationship defined her condition as puerperal insanity. By the twentieth century, changes in all three layers made the disease obsolete. The creation and demise of puerperal insanity illustrates not only the social construction of illness but also the cultural embeddedness of medical categories.

Introduction to Chapter 5

Theriot's essay, itself a sophisticated contribution to its own field of interest, serves also as a bridge to the age of hierarchy in the later nineteenth and early twentieth centuries. In his suggestive essay about James Emerson, Edwin T. Layton has revealed the notions and actions of a vanquished faction of inventors in one of the age's most terrific cultural struggles, between champions of early and mid-nineteenth-century democratic folk culture on the one hand and those of late nineteenth- and twentieth-century expert culture on the other. The professionalization of science and the emergence of so-called science-based technology in that era undermined the position of the independent craftsman or mechanic, that celebrated folk hero of American culture. James Emerson fought for the tinkerer, who typically made so-called emulative devices, that is, relatively small improvements on preexisting devices without the assistance of special scientific knowledge. Indeed, the invention for which Emerson was best known was a trivial example of the craft-based invention of emulation, the mustache cup. Its purpose was to keep facial hair out of a beverage cup; Emerson's idea was to install a comb across the cup's mouth, definitely a small, commonsensical improvement on an existing device. Clearly this was the independent mechanic's turf.

Born in New Hampshire, Emerson was a self-taught mechanic and inventor. He worked at a variety of trades, including that of millwright. In the later 1860s he was invited to use his relatively crude dynamometer at Lowell, Massachusetts, to measure power and efficiency in water turbines; throughout the 1870s he tested the claims of turbine manufacturers as to the virtues of their products, chiefly for the water company of Holyoke, in western Massachusetts. In 1880 he was replaced by an engineer, even though his measurements had had a rough-and-ready utility and accuracy. In sum, he could not compete with the credentials and prestige of the new scientific engineers. Layton suggests several reasons for Emerson's dismissal, including his radical politics and the superior precision of the engineers' devices and methods. Emerson went to contest the new breed of scientific engineers' claims to precise measurement of the power and efficiency of the region's industrial water turbines. Precise measurement was the scientific engineers' substitute for the rule-of-thumb methods that independent mechanics such as Emerson used, and the battle was symptomatic of the struggle, not over technical knowledge and its utility as a general proposition in industry and technology, but over *whose* technical knowledge, and *which* technical community—the new scientifically trained engineers or the traditional craftsmen—would triumph. Emerson's strategy was to offer what Layton dubs a reasonable measurement for far less money in this contest, appealing all the time to the old-time virtues of democratic craftsmanship and knowledge.

Emerson thus wished to advance the interests of the entire class of inventors to which he belonged, the traditional craft-oriented mechanics. Layton perceptively draws parallels between the independent inventor, or the craftsman-mechanic, and the yeoman farmer, yet another American cultural hero. He argues further that Emerson's populist technology was a kind of cultural construct as myth, for clearly he, as a craftsman-inventor, was fighting a rearguard action in his own time on behalf of his millwright colleagues. Of course his argument that craft-based inventors could still meet the nation's technical needs was undercut by his own adoption of some of the new scientific techniques. Here indeed was a battle royal, at least symbolically, over what kind of technical knowledge mattered and whom it benefited and whom it did not. Layton insists that the importance of the craftsman or mechanic was increasingly a myth, although it was one that Emerson continued to invent or even reinvent. Emerson railed against those important cultural constructs of the late nineteenth and early twentieth centuries, hierarchy, centralization, standardization, and, above all, expertise.

EDWIN T. LAYTON

The Inventor of the Mustache Cup: James Emerson and Populist Technology, 1870–1900

James Emerson was one of the inventors of the mustache cup. It was a trivial example of an important class of inventions. Brooke Hindle has reminded us of a craft-based type of invention that Adam Smith and other eighteenth-century figures categorized as "emulation."[1] For our purposes "emulation" refers to relatively small improvements in existing artifacts based on craft knowledge, common sense, and experience but without any significant aid from science. Though Emerson did not use the term, he saw inventions of this type as a democratic alternative to science-based invention and the professional elitism in technology that came with basing technology upon esoteric scientific knowledge. Emerson became a rebel who sought, in a manner reminiscent of contemporary populists, to restore a more democratic America and who emerged as the champion of the traditional self-employed craftsman and a democratic technology.

America as a democratic society had long fostered democratic and egalitarian tendencies in its science and technology. In particular, Americans looked to independent inventors, usually people of humble origins, to provide their technological innovations. The period from the adoption of the Constitution until the start of the twentieth century was a golden age for individual inventors. The self-trained engineer and inventor long had few rivals in the United States; professional engineers were slow to develop in America and did not become the dominant group until the end of the nineteenth century with the rise of big business structured by corporate capitalism.[2]

In the modern age, invention has become scientifically based and associated with giant corporations or government agencies. While individual inventors have by no means been eliminated, technological change has come to be associated with bureaucratized science and technology in the employ of large organizations, particularly private corporations. The need for esoteric knowledge has produced technical elites of scientists and engineers who displaced inventors and craftsmen from their position of leadership in technological invention and innovation. Corporations sought to systematize the process of technological change, as for example in the modern industrial research laboratory.[3] One part of this drive for rationalization and systematization involved substituting engineers, increasingly college-educated engineers, for mechanics and craftsmen in carrying out significant, creative technical work.[4]

The rise of professional science and engineering elites eroded the social as well as the technical position of craftsmen, as of farmers. The case of farmers is especially striking. To Jefferson and Jackson, farmers were the moral, political, and economic backbone of the republic. But this view of farmers was challenged by the emergent agricultural scientists, such as Harvey Wiley. Wiley saw the farmers as a dependent class, a problem group whose welfare required the leadership of a scientific elite. As Alan Marcus has shown, the agricultural scientists led a movement that would eventually wrest technological and scientific leadership in agriculture from the more advanced farmers. The crisis for farmers was not merely loss of leadership; they resented the loss of their traditional self-image as autonomous repositories of wisdom and virtue and the reduction of their role to that of dependant laborers.[5] The loss of power, autonomy, and self-image no doubt contributed a cutting edge to the many movements of farm protest and reform that culminated in the populist revolt. Farmers moved inexorably from independent citizens in the nineteenth century to a class dependent upon the government and private corporations for the science and technology they needed, and upon a powerful political lobby to ensure government subsidies in the twentieth century.

The situation of craftsmen was similar to that of farmers. During the Jacksonian era mechanics had stood with farmers as independent pillars of a democratic community. Later in the century, they lost their position as pillars of a democratic society and technological leaders of the republic. Instead they became a sometimes troublesome class, hourly employees with a tendency to form craft labor unions and strike. Craftsmen retained only a modest, secondary role in technological change. By the twentieth century, craftsmen were becoming part of the growing "labor problem," just as farmers were becoming part of a growing "farm problem."

The loss of leadership by independent inventor-craftsmen was obscured, and the transition was buffered by an ideological myth of the lone heroic inventor. Thomas A. Edison, for example, did much to further the myth that nothing had changed. However, as Matthew Josephson and others have pointed out, Edison was scientifically informed. Edison exploited the myth of the heroic inventor for public relations purposes, obscuring his greatest invention, the multidisciplinary industrial research laboratory.[6]

Few engineers were deluded by the continuing myth of the heroic, non-scientific inventor. They knew that technological leadership was shifting to scientifically based practitioners like themselves. Engineers and scientists were becoming increasingly professionalized and self-conscious, and they made rather sweeping claims for the role of the scientific elite. George S. Morison, a leading civil engineer, in 1895 waxed lyrical in articulating the glorious mission of engineers as the leaders of a new civilization. As Morison claimed: "We are the priests of material development, of the work which enables other men to enjoy the fruits of the great sources of power in Nature, and of the power of mind over matter. We are the priests of the new epoch, without superstitions."[7]

The conflict between craftsmen and corporate engineers began with the rise of the first important industrial corporations in America, those founded by Francis Cabot Lowell and his Boston associates to produce textiles in New England. As John Kasson has argued, on one level the Waltham system of labor and the sorts of communities developed at Waltham and later on a larger scale at the industrial city of Lowell were adaptations of British textile technology to American republican values and ideology.[8] But on another level, the textile corporations at Lowell and other New England textile centers presented a challenge to the independent, self-employed craftsman and inventor. The initial technical leadership was provided by nonprofessional technologists: Francis Cabot Lowell, a capitalist with technological vision, who was aided by a craftsman-inventor of genius, Paul Moody, in developing the power loom and other elements of a novel reworking of British textile technology that emphasized labor savings.[9]

After the initial work by Lowell and Moody, leadership shifted from craftsmen to professional engineers. In many ways the founding father of New England hydraulic engineering was Charles S. Storrow. Storrow received a secondary education in France, and after his graduation from Harvard he returned to France to study engineering.[10] He translated a number of French scientific works on hydraulics into English, thus helping to provide a basis in esoteric knowledge for hydraulic engineering practice. Another Boston engineer translated a standard French text on hydraulic engineering into English.[11] French engineering placed great emphasis upon science, especially mathematical theory. Storrow and his followers emphasized experimental science and were less theoretically oriented than French contemporaries.

Storrow and two other prominent engineers were employed by the proprietors of Lowell to do important scientifically based engineering work, particularly in measuring the flow of water in open channels. He then went on to become one of the principal architects of the new industrial city of Lawrence, ten miles downstream from Lowell.

Storrow's scientific approach to engineering problems was continued by his successors at Lowell, notably Uriah Boyden and James B. Francis. Boyden and Francis were thoroughly conversant with the international scientific literature bearing upon hydraulic engineering. Both Boyden and Francis sought, like Storrow, to base engineering practice on science. They differed from French college-educated contemporaries mainly in their emphasis upon experimental science and their skepticism of purely theoretical reasoning in technology.[12]

Perhaps nowhere was the scientific leadership of hydraulic engineers shown more dramatically than in the shift from the craft-based technology of vertical water wheels to the scientifically based technology of the hydraulic turbine. In the early days of Lowell, Paul Moody, whose initial training had been as a millwright, installed large vertical water wheels to power the textile mills. These traditional vertical water wheels were the products of centuries of craft evolution; they were relatively inefficient, and they converted at best only about two-thirds of the power of the falling water into useful work. Uriah Boyden was a pioneer in introducing the turbine of Benoit Fourneyron into New England. Boyden made a number of important improvements in Fourneyron's turbine, which, under favorable circumstances, was able to convert as much as eighty-eight percent of the power of the falling water into useful work. Boyden also was an important pioneer in improving methods of testing turbines to determine their power and efficiency by use of the Prony dynamometer. Francis made contributions to every department of hydraulic engineering, gaining international fame for his invention of a new type of turbine, the Francis turbine, and his further improvements in the methods of testing turbines.[13]

The work of engineers such as Francis lay, above all, in replacing rule-of-thumb craft methods by precise, scientific measurement and quantification. Perhaps nowhere was this more apparent than in Francis's improvements in methods of testing turbines. The basic device tool, the Prony dynamometer, had been invented in France by an engineer, Baron Gaspard F. C. Riche de Prony (1755–1839), a pioneer in scientific engineering and one of the first professors at the Ecole Polytechnique. Prony's dynamometer was one of the tools at the cutting edge of scientific technology. It was intended to replace guesswork by exact science in measurements of the power output of prime movers. Only by exact measurements could the benefits of an improved design be clearly proved in a verifiable way. Prony's dynamometer operated by a massive friction brake. The brake was tightened around an iron wheel mounted on the shaft of the water wheel or other prime mover. The output of rotary power (torque) was absorbed as friction on the brake. The pressure on the brake needed to extract this power gave a measure of the power output. Francis not only improved the use of the Prony dynamometer (which measured power output), but he engaged in extensive experiments to derive a formula for the amount of water used (the input). Francis's "weir formula" was a critical contribution that made truly accurate measurements of input and efficiency possible.[14]

The Prony dynamometer is typical of the tools of precise quantitative measurement that engineers created in every department of technology. Not untypically, it did not work very well at first. The massive friction produced vibration that made precise measurement difficult. To calculate the efficiency of the turbine—that is, the relation of input to output—engineers had relied upon formulas to estimate the input, the amount of water flowing through the turbine in a given period of time. These early theoretical formulas were inaccurate. Francis derived a formula that gave much better results and that is still the basis of measurements of the flow of water over weirs. Francis also overcame the problems of precise measurement using the Prony dynamometer. Building on the work of his colleague and associate, Uriah A. Boyden, Francis resorted to a number of expedients to increase accuracy. By painstaking, exhaustive, and expensive efforts, Francis succeeded in making the testing of turbines into something approaching an exact science.[15]

The contrast between Francis's exact measurements and the approximations used by millwrights was dramatic. It was the difference between scientific precision and rule of thumb. Only by challenging the professional engineers' preeminence in measurement could craftsmen hope to regain their old position. This was precisely what James Emerson did. Emerson invented an improvement of the Prony dynamometer. Francis got accurate results by employing many observers and taking other labor-intensive measures. The cost was high, however; it cost about twelve hundred dollars and sometimes more to test one turbine using Francis's methods. Emerson cut time and costs by using a more compact iron friction brake that was hollowed out and cooled by water. This reduced the heat and vibration, making it possible for a single observer to make measurements of reasonable accuracy. Emerson used one assistant in making the tests; Francis used five or more. Emerson left out many of the refinements used by Francis; he did only a small number of the experimental measurements in a matter of a few hours, thus saving much money. Francis did multiple experiments, usually over several days.

Emerson got his start at Lowell in 1868–69 when he was invited to use his dynamometer for public tests of the turbine of Asa Swain, a millwright-inventor. Emerson built a testing flume at Lowell to conduct these tests. He showed his flair for publicity in making a success of his testing flume as a continuing enterprise. Turbine companies published catalogs in which they set forth the virtues of their products. Their claims to efficiency were, however, dubious in many cases. Emerson made the public aware of how doubtful such unverified claims were. The success of his flume at Lowell from 1869 to 1874 caused the Holyoke Water Company to invite him to come to Holyoke in 1874 and manage a larger turbine testing flume and operate it for the benefit of the consumers of water power. Emerson operated the Holyoke Testing Flume from 1874 to 1880, when he was removed and replaced by a distinguished hydraulic engineer, Clemens Herschel. Emerson was replaced because his measurements were too hasty and crude, though they had proved useful all the same. It is possible that his radical social and political views may have played a role in his removal.

The work of the Lowell and Holyoke testing flumes was of considerable importance. It served in something of the role of an industrial standard setter. Emerson succeeded in part because of relentless publicity. He published a periodical, *Emerson's Turbine Reporter,* which printed the test results along with Emerson's spicy, iconoclastic social comments and political philosophy. He also published letters to the editor and his own replies. Emerson initially published five thousand copies of each issue, but his feisty social commentary and his discussions of spiritualism increased demand and brought him tidy profits. Emerson collected his test results, along with essays on industrial topics and colorful social commentary in a book, nominally on turbine testing, but including much social and political musing. Emerson's book went though six editions and appears to have reached a wide readership.[16]

Emerson's work at the Lowell and Holyoke testing flumes was important. There can be no doubt that it played a significant role in the advancement in performance of American-made turbines, particularly improvements in the Francis turbine made by craftsmen. These improvements were the basis for Emerson's claims that craft-based inventors could still meet the needs of the nation and that college-educated engineers were not needed. And although Emerson's claims for the ordinary mechanic and self-trained inventor were exaggerated, they were not without significant basis in fact. Emerson did make turbine tests available to all comers at low price. These tests became a substantial source for subsequent improvements in the turbine. As Emerson claimed: "Nine-tenths of the water wheels brought to him that first year only gave three-fourths of the power which their builders claimed. . . . The influence of these texts was beneficial to every honest builder. The first wheel tested by the Stillwell and Bierce Manufacturing Co. only gave 68 percent [efficiency], although they honestly believed it could be relied upon to give 85 [percent]. When they discovered the truth they commenced experimenting and improving their wheels until they gained records of over 90 percent. A similar improvement was made by [other leading turbine makers]."[17]

Emerson used the Francis weir formula as an integral part of his turbine testing. This was an awkward fact for Emerson, since the weir formula was a clear example of the use of science in engineering. Emerson granted (with evident reluctance) that the Francis weir formula was excellent, but he attempted to diminish its obvious importance by suggesting that changing the proportions of the experimental setup "renders the formula worthless."[18] It was, however, far from worthless to Emerson. Emerson had his mathematical assistant use Francis's formula to calculate what Emerson called the "Emerson Weir Tables," which took sixty-one pages of fine print and involved twenty thousand calculations.[19] They served to make the data available in advance in the form of handy tables, so that the sometimes laborious calculations did not have to be performed for each test. By incorporating Francis's greatest scientific contribution to engineering in a set of tables to which Emerson attached his own name, he sought to minimize his dependence upon Francis and upon science. In dealing with an important refinement introduced by Francis, Emerson merely said that it was "the method adopted at Lowell,"[20] without specific reference to Francis.

Emerson's borrowings from Francis and his science were crucial to Emerson's success, but he was almost pathological in his denunciations of Francis as an engineer. Francis was not just a very distinguished engineer; he also became a role model for the professionalization of engineers. He was, in short, the symbol of all that Emerson hated. Francis may have made the pugnacious Emerson even angrier by his constant courtesy and his apparent obliviousness to Emerson's attacks. Emerson held that "few men have had Mr. Francis's opportunities, and I think few would have given the world so little in return."[21] He thought Francis as engineer was wasteful and overly theoretical. He alleged that Francis "will give opinions from books very readily, but would start back with as much haste as a young maiden does from a proffered kiss, if one should present a petition for his signature for anything except the interest of the Lowell companies."[22] Emerson's venom may have been due to his unavoidable dependence upon Francis's work and the fact that Emerson had put the qualifier, "while acknowledging many kindnesses from Mr. Francis" before his conclusion that, "I believe him to be a hindrance, so far as his influence goes, to perfection of the turbine and improvements generally."[23]

Emerson linked his attacks on Francis with his denunciations of college education of engineers. Though Francis had not attended an engineering college, Emerson saw (correctly) that Francis was the model that engineering colleges aimed at in shaping their curricula. Emerson denied the value of college education for engineers: "Of the hundreds of young men who yearly graduate from our educational institutions, how few of them are ever likely to reflect credit upon the name, simply because nature never intended them for the business."[24] To Emerson the essence of engineering was ingenuity, which he held was an inborn trait best developed by practical work. He argued that the necessary qualities of ingenuity, judgment, and character were developed by practical work, as he thought had been the case with Watt, Fulton, and Stevenson. In other words engineering genius was an innate propensity brought out by practical experience.

To Emerson the success of his testing flume and the improvements that took place in American turbines vindicated his ideas about technology. Emerson claimed that the turbine improvements to which he contributed were entirely independent of science. He tried to get statements from leading turbine designers to support his position. In this he failed, but he was honest enough to publish the results. He elicited comments for the *Reporter* from three of the leading turbine designers of the day, T. H. Risdon, Asa M. Swain, and John B. McCormick, but this effort backfired. Both Risdon and Swain wrote letters indicating the nature of their design methods; in both cases they were scientific. Swain showed an awareness of the European literature on turbine theory, to which Emerson felt compelled to reply. Turbines were subject to "the simplest measurements, rendering it useless to theorize about" them, Emerson maintained.[25] Risdon presented an original theory of design that was scientific in spirit, starting with the premise that turbines worked by reaction. Emerson did not comment on Risdon's letter, though he kept insisting that leading turbine designers did not use scientific methods.[26] Emerson was somewhat more successful with John B. McCormick, another creative designer. Emerson avoided asking McCormick about science, but inquired only whether he had derived benefit from the old textbooks. McCormick replied that old textbooks had been of no benefit to him in his work.[27] It is not clear how McCormick interpreted the phrase "old textbooks," but undoubtedly McCormick drew upon a rich tradition of vernacular science in his creative work.[28]

Emerson believed that creative inventions depended on ingenuity, common sense, and experience; he denied any role to science. Emerson lost no opportunity to denounce science and education for engineers. In replying to one of his letters (signed "Millwright") Emerson denounced all science. He argued that "Millwright's" letter was "highly charged with old terms, ideas and obsolete theories. . . . so with the theoretical waterwheel builder;

he can hardly speak without letting out an impact, radial vanes, gravity, vertical lines, direct action, re-action, periphery, co-efficient, radial floats, etc." Emerson argued that such (scientific) terms were of no benefit and that they were not used by successful turbine designers.

Emerson's denial of a creative role to science in technology was not wholly false. Clearly, the turbine designers did use theory and science, but the perfection of a new design involved a great deal of "cut-and-try" of the sort that Emerson advocated. The turbine makers were, of course, intimately familiar with the workings of their products. Once manufacturers were aware of a problem, their knowledge of the state of the art and their experience would suggest possible remedies. Emerson's testing flume allowed them to try these variations one at a time, until they were able to get satisfactory performance. Tests often extended for a period of many months, during which the makers were allowed to make alterations and improvements. Emerson did not publicize the unsatisfactory test results, only the successful final tests. In effect, Emerson's testing flume became a development laboratory in which products could be debugged, sometimes after many months of tinkering and dozens of tests. Such "cut-and-try" empirical methods are quite powerful and effective.[29] Emerson never considered the possibility that such methods would be more powerful if the turbine builders were theoretically informed, so that the "cuts" were not random guesses but rational judgments based on theory as well as experience and practice. And this was indeed the case. Thus the success of his testing flume was not a vindication of Emerson's belief that craft technology without science was sufficient for the nation's needs.

Emerson's "improvements" on Francis's methods of turbine testing did not make them more accurate. On the contrary he sacrificed precision to get low cost. Emerson's tests were quick and dirty and lacked the painstaking determination to achieve the highest accuracy that characterized Francis's work. But they were cheap. Emerson stressed the fact that testing as practiced by Boyden and Francis made assessment "so expensive as to be beyond the reach of any but wealthy corporations."[30] Emerson's lower costs were important because they allowed such testing to become part of the development of new turbine designs. Thus Emerson made a seemingly scientific technology into one amenable to traditional craft methods. Emerson was not exactly modest; he was convinced that his cheap turbine testing had revolutionized both the science and technology involved. He claimed: "I as hydraulic engineer have done more in the last quarter century to establish a practical knowledge of hydro dynamics . . . than has been rendered by all other engineers for a century past."[31]

His publications made Emerson widely known among millwrights and associated mechanics. To many he became a hero who offered to restore the craftsman to a position of technological leadership. Emerson's social commentary amounted to the formulation of a populist theory of technology and invention on the one hand and a populist indictment of elitist tendencies in American society on the other. To understand this populist revolt of craftsmen, we need to look in more detail at Emerson's own background and work.

Emerson published a colorful autobiography in his *Turbine Reporter.* He was born on a farm in New Hampshire. "I had an intense desire to plow the mighty deep, though none to plow the rocky soil of New Hampshire," he wrote, and like many poor New England lads he went to sea. He signed up as a seaman on a four-year Pacific cruise of a New Bedford whaler. He rounded the Horn and learned the sailor's life. He resented the discipline and jumped ship. Emerson was imprisoned in a jail in Chile, and after a series of colorful adventures (of possibly doubtful authenticity), he made his way back home to New England. In his autobiography he cast himself as the underdog ever resisting the tyranny of the powers that be. He had a mechanical turn of mind and became an inventor of some ability. Though he never attended school, he picked up a literary and scientific education of sorts by self-education and reading.[32]

Emerson settled in East Boston and worked at mechanical pursuits after his return to the United States. He invented a novel ship's windlass that he struggled to market from 1850 to 1860. Emerson's idea was to mount the windlass on massive reduction gearing made of cast iron. (The windlass was also made of iron later on, if not from the start.) He saw that gearing would provide more mechanical advantage in lifting heavy weights and that cheap cast iron gearing would be sufficiently strong if it was made massive (cast iron gearing had been extensively used in early textile factories). His first insight had come to him in 1850, but when he sketched his idea of using massive cast iron gearing to some of his "seafaring friends" they reacted negatively. Emerson quoted them as saying, "What! Trust lives and . . . property to cast iron gears? might as well have a glass windlass."[33] (Cast iron is brittle, but if sufficiently massive it has considerable strength.) Once he perfected his invention he encountered resistance and had to struggle for recognition. As Emerson described it, "The windlass was so radically different from all previous devices for the purpose, that it was laughed at by seafaring men, particularly naval officers, etc. Four years of persistent effort and a gift to an impecunious ship-owner gained the privilege of putting one on a ship."[34] Once in use Emerson's geared windlass proved itself in several years, and it found many applications during the Civil War.

Emerson formed a company to manufacture his windlass in 1857. After many struggles, he convinced buyers of the utility of his device. He gradually gained a modest success and sold his patents and business to a corporation in 1860. Emerson claimed that the successor company in the United States installed more than two thousand of Emerson's geared windlasses on American ships, and its British affiliate allegedly put an additional six thousand of Emerson's windlasses into operation. It is possible that its success was in part due to the fact that it was a transitional device. The windlass part could be operated manually, but the geared hoisting apparatus could either be worked by a steam engine or connected to any external power source.[35] Emerson appears to have derived a modest amount of money from the sale and modest royalties subsequently.[36]

Emerson's windlass, a direct outgrowth of his maritime experience, was probably his first significant invention. But even before the success of his windlass in 1856, Emerson turned his attention to factory machinery. In 1856 he received his first important recognition as an inventor; the Massachusetts Charitable Mechanics Association awarded his windlass a gold medal. But at the same time the association also honored another of his inventions with a gold medal, his "power scale." This was a simple dynamometer for measuring the power consumed by a machine by measuring the power with a compact dynamometer mounted on the shaft from which the machine got its power. Its practical value was that it measured the power used by particular machines in a factory. Emerson later published tables of his tests of various kinds of machinery. These were in the nature of examples; Emerson realized that particular circumstances determined the power needed, and the tests had no absolute value. The same machine might take more or less power in different settings. The utility of Emerson's device was that it allowed comparisons of the power usage of different makes of machines used in the same factory.[37] It probably provided him with the idea of developing an improvement in the Prony dynamometer and a career in turbine testing.

All of Emerson's inventions fit the pattern of "emulation." Emulation was a craft-based style of invention, which involved improving upon the work of others.[38] All of Emerson's inventions appear to have been technically modest improvements in existing devices. However, it is important to realize that the economic and social impact of an invention bears no obvious relation to its technical sophistication. In the case of the geared windlass Emerson's improvements were economically important. His improvement of the dynamometer was of both social and economic importance. Emerson also patented a system for heating railroad cars with steam from the locomotive. The same general idea occurred to other inventors at about the same time; Emerson's apparatus was tested but not adopted by the railroads. He had no better luck with a steam brake and an automatic coupler that he invented and tried to sell to railroads. None of his railroad inventions appear to have been successful.[39]

Emerson's best known invention was the mustache cup. Though Emerson professed to be ashamed of this invention, this device will, nevertheless, help clarify both the strengths and weaknesses of Emerson's inventive style. To quote him: "A lady friend as a joke asked me to get up a device to keep her husband's mustache out of his coffee. A plan was readily found consisting of a peculiarly shaped comb. . . . Two young ladies asked to have a patent taken out and assigned to them. It was applied for. . . . the man for whom the plan was devised . . . exhibited it; in less than a month several hundred dollars worth of orders were received from fancy goods dealers."[40]

Emerson claimed that he thought the joke had gone stale and that he had abandoned it in disgust. His high-minded disdain may have reflected his inability to collect royalties on this invention. Emerson did not blush to promote his "Duplex Piano Stool" (the patent for which he did own); it was technically on a par with his mustache cup.[41] Despite Emerson's disgust, the mustache cup was typical of emulation. It was an improvement on an existing device, which relied upon practical ingenuity applied to a common article. Like most such "emulations," it was capable of further improvement by others: a solid masking rim is simpler and better than the comb used in Emerson's mustache cup. It was easy for anyone to understand, because all are familiar with the act of drinking from a cup. He showed ingenuity, but nothing in his invention was new; he combined elements that were already familiar (in this case a comb and a cup).

Emerson's other inventions were similar to his mustache cup, except that they assumed familiarity with machinery such as geared lifting engines and dynamometers. In principle his improvements of the windlass and dynamometer were not different from his invention of the mustache cup. He drew upon a native ingenuity in order to improve an existing device by methods or components already well-known in other contexts. In both cases Emerson took a device hitherto made of wood and made it of iron, a commonplace (but important) change in material. Internal cooling and gear trains were scarcely new; to mechanics and millwrights they were as familiar as cups and combs were to ordinary unmechanical people.

Though the mustache cup model of invention seems trivial, it is quite powerful, as Emerson demonstrated with his dynamometer. Emerson showed strategic insight into the weak spots of older Prony dynamometers. By using metal instead of wood he could reduce the size and weight of the friction brake. By having his metal brake hollow he could pump water through it to absorb the waste heat produced by the enormous friction inherent in this type of dynamometer. Iron, a much better heat conductor than wood, lent itself to internal cooling. The rigid metal construction reduced the intense oscillations that made precise measurement such a problem with the original Prony dynamometer. Emerson's testing flumes involved both of his major inventions. By means of his windlass he could lift a turbine into place in his flume and remove it after the testing was done. By means of his friction brake he could make the critical measurements needed quickly and cheaply.

A millwrights' revolt against science coalesced around James Emerson. No doubt many craftsmen had been unhappy for some time with scientific analyses that they could not understand. Their unhappiness must have been made more acute by the devaluation of their own craft knowledge and the threat of technological obsolescence that they faced from the growth in the number of college-trained engineers and of a body of scientific knowledge for them to use. The 1880s and 1890s were an age of agrarian protest. To country millwrights, protest and rebellion against authoritarian ideas supported by an eastern establishment appeared to gain a new legitimacy.

A critical element in the rejection of science by millwrights was their rejection of reaction (Newton's third law of motion). To them a simple, intuitive idea of pressure acting directly was all that was needed. The antiscientific millwrights coalesced around the idea of "direct action," code words for rejecting science. One of most outspoken millwrights, T. W. Graham of Dubuque, Iowa, maintained, "The division of the weight of falling water into two forces is one of the grossest absurdities which still linger in the text books."[42] An assertion of the role of reaction in turbines in 1882 led to a flurry of letters to the editor of *American Miller* from the antiscientific advocates of direct action. This flurry was repeated in the magazine, *Milling*, in 1893. Advocates of direct action explained reaction as a special case of action. In a rotor such as Barker's mill, the apparent force of reaction was no more than an illusion, caused by the unbalanced pressure created by the outlet. In 1883 William Kennedy in defending the unbalanced pressure interpretation of reaction referred to "us direct action fellows."[43]

In rejecting science, the millwrights found a substitute in the empirical data produced cheaply by Emerson's dynamometer. In 1893 Graham pointed out that in one of the areas in which the most dramatic progress had been made in turbine design, the efficiency at part gate opening, "the dynamometer and the testing flume are the agencies which have solved the problem of partial gateage, rendering it possible to build turbines which will develop from seventy to seventy-five per cent when discharging half the water they discharge at full gate. . . . twenty years ago such results were unthought of."[44] This was then a case where empirical improvement, that is, emulation, had apparently been able to produce improvements that were not the result of increased scientific understanding. Such cases were telling for Emerson and the craftsmen who resisted science; they seemed to show that craftsmen using common sense and empirical measurement could make improvements without recourse to science and engineering. It is significant that the mechanical engineer C. R. Tompkins, who opposed the unscientific theorizing of Graham and other advocates of "direct action," also criticized Emerson's test results, which he though were inaccurate and overstated the efficiency of the turbines tested.[45] But however skeptical an engineer like Tompkins might be, he acknowledged that "the merits and demerits of the celebrated Holyoke testing flume has had its full share in these discussions."[46]

The millwright revolt was at heart populist. Tompkins, the engineer, confronted Graham with the absurdity of conflating force of impact of a stream of water from a hose with the backward force exerted by the hose on the hand holding it.[47] Graham's response revealed the populist overtones to the antiscientific revolt: he maintained that "only when such litter and rubbish is swept from the textbooks, will hydraulic principles be better understood by the masses."[48]

Emerson's social and political commentary published in his *Reporter* and the later editions of his books contained a populist critique of the status quo. Most of his political and some of his social criticism was consistent with that adopted by the populists. Emerson never mentioned the populists whose strength lay in the rural West, not the industrial East. It is doubtful that the populists would have wanted to claim Emerson, despite the similarities in some areas, because Emerson's denunciations of Christianity and his support for free love went far beyond the political platform of the Populist Party.[49] Emerson's linkage of Christianity with the moneyed interests, however, would have struck sympathetic chords in some radical social critics.[50] But he showed only a generalized sympathy for the farmers and did not address the issues, such as railroad rate abuses and currency inflation, that were at the heart of the populist program. Emerson was a "populist" without the capital P.

Some of Emerson's most important statements were made in cartoons or drawings that required little literacy to understand. Emerson was very clear about his social ideal: it was the independent mechanic. While Emerson often upheld the primacy of the craftsman, no statement of his was more effective than his cartoon, "The Mechanic to the Front." He pictured a handsome young couple and a shepherd caring for sheep but with symbols of progress: the steam engine and the telegraph. The caption noted that "the real creative mechanic approaches nearest to the attributes of his creator," and "steam, the telegraph, telephone, electricity all are in their infancy and sure to develop new fields of progress."[51] The implication of this (and many other statements) was that theory and college professors were of slight value and that future technological advances would come through the ingenuity of practical mechanics. That is, Emerson wanted to restore the craftsmen to their former position of technological leadership. Emerson showed a young couple, the mechanic and his wife, and he clearly appealed to democratic social norms and egalitarian technological traditions in this and other drawings and writings. Though his hero was the mechanic, Emerson the farm boy had a general sympathy for the farmer, for example in the comment, "The unpretentious farmer that place a watering trough by the wayside for the thirsty man or horse, in my opinion, does more for the elevation of man and the glory of God than the rich man who builds a church or endows a college."[52]

Conversely, Emerson thought education a waste of time. He was particularly outspoken in denouncing college training of engineers as a positive hindrance to creative engineering work. But he was equally opposed to all formal education, particularly college education. On a drawing of a college he placed the caption, "Show Institutions of the Hills, for the Blind, the Halt, the Idiotic, the Insane, the Pauper, and the Criminal,"[53] which he portrayed in his cartoons as promoting social elitism and old boy networks.

In one of his editorials Emerson denounced Tyndall and other scientists. Under the title "Scientists and Professors," Emerson argued that with time the works of scientists (such as Tyndall) would be forgotten but that great mechanics and inventors would remain folk heroes, and he cited names such as "Franklin, Morse, Fulton, Watt, Howe." All had been, of course, self-educated. But scientists and engineers were not his only targets. Emerson denounced physicians for their blundering and guesswork.[54]

Second only to the clergy, Emerson relished denouncing the courts and the legal system. In editorials and in his cartoon "The Law," he compared the court system to open privies discharging on the railroad's right of way. Both were equally "progressive," Emerson held. Emerson pictured the witness box of the Massachusetts court as a miniature gallows with a noose, reflecting no doubt the many defeats he had experienced at the hands of engineers as expert witnesses.[55] Under the title "Rotten Statutes," Emerson waxed sarcastic, "how we smile at the Jay Goulding of a railroad through the chicanery of the law. . . . If the law is designed to aid justice why is such rascality tolerated by people claiming to be civilized?"[56] He titled another essay, "The Law Antagonistic to Knowledge and Justice."[57] But in his tirades against injustice, Emerson showed no concern for hourly laborers, nor for the railroad rate abuses and other injustices of which farmers complained. (His chief complaint concerning railroad management was its lack of concern for the passengers' comfort, a problem that could be remedied by adopting his patented system for steam heating railroad cars.[58]) His focus was the independent mechanic, so his bitterest complaints were against the patent system and all those things—corporations, elites, scientists, engineers, and colleges—that served as barriers to the mechanic regaining his traditional place in society.

Emerson's attacks on the Republican Party and corrupt politicians were surprisingly mild. He denounced the party's loss of its earlier idealism, its corruptions, carpetbag governments, and favoritism shown monopolies. But this short essay was less than a page in length and squeezed between essays on "Adulterations and Short Measure" and "The North Pole."[59] Emerson also denounced the protective tariff (in a short paragraph between "Lisle Thread" and "Fire Escapes.")[60] Except for the patent law, Emerson did not see government as either the cause of or the cure for the evils with which he was most concerned.

Emerson's populist technology was a failure. He did not succeed in his grand objective, to restore the craftsman to a position of technological leadership. Nor did he halt the trend toward education and the appearance of scientific, technological, and professional elites. But he did call attention to both issues and came up with plausible but ultimately futile remedies. Emerson's work was not entirely wasted, however. The principle of public testing contributed to a body of public knowledge about technology that was important, perhaps ironically, for the development of a scientific base in technology. Emerson's tests may have been slightly crude, but they were useful and facilitated the recognition of the necessity of a tedious process of development for complex technological systems. Indeed, Emerson's revolt had the ironic result of increasing the power and influence of science-based technology and the elites associated with it.

One of Emerson's innovations in particular was critical to the development of a more sophisticated science of the turbine than that which existed in his day. European turbine makers produced only "one of a kind" turbines tailored to the needs of individual users. In the United States the larger firms produced a whole range of sizes of geometrically similar turbines from a diameter of about fifteen to eighteen inches up to large machines of forty-eight or fifty-two inches in diameter, which they listed in sales catalogs. Emerson was able to persuade several leading turbine manufacturers that they needed to test not just one size but all of their sizes (since performance could and did vary unpredictably as the scale changed). He had had some success by 1880 when the Holyoke company removed him and replaced him with Clemens Herschel. For example, Emerson made a series of tests of virtually the entire line of turbines produced by the firm of Stout,

Mills, and Temple, of Dayton, Ohio, the second largest American producer.[61] Herschel continued Emerson's practice and extended it. The existence of these results, giving performance data for geometrically similar turbines of different size and at constant head, provided material that was used by Rudolph Camerer in his development of a new and vastly more sophisticated theory of the turbine, based on his discovery of two key dimensionless parameters that emerged from the data acquired by Emerson and Herschel.[62] (These were the "specific speed" and the "capacity coefficient.") The net result of this was to make turbine design the exclusive province of scientifically grounded engineers, thus excluding in the end craftsmen-inventors such as Emerson, Swain, and McCormick. Emerson ended by advancing the very sort of scientific technology and technical elites that he detested. The issues of the dignity and autonomy of the technologist that Emerson raised did not, however, go away. They were reborn in the corporate context of modern engineering in the twentieth century. They remain valid and important issues for today's world.

Introduction to Chapter 6

The role of technical knowledge in American cultural constructions such as hierarchy, centralization, and expertise figure prominently in Zane L. Miller's essay, an arresting analysis of that most formidable opponent of racial bigotry, the African-American social scientist William E. B. Du Bois. Miller argues that Du Bois thought in group or racial categories, just as did virtually all his fellow citizens, including most emphatically his opponents in the racial conflicts of the early twentieth century. There was, to be sure, a crucial difference. Du Bois did not accept the fatalistic belief of most of his social science colleagues—and most Americans, for that matter—that individuals could never transcend the limitations (or opportunities) of the group identity that they shared in common because of the dictates of a deterministic nature and nurture. And Du Bois agreed with that other lonely social scientific voice on racial equipotentiality and indeterminancy, the Columbia University anthropologist Franz Boas. Du Bois did not accept the all but universal belief among Americans that there was a natural and permanent hierarchy of races, based on their skills, talents, and expertise. Du Bois agreed with Boas that groups did exist in the contemporary social order. All of this could change, however. Unlike Boas, Du Bois believed that the eight races of mankind each had their own "genius," which enabled or would enable them to adapt and contribute to modern civilization. Thus Du Bois thought in terms of racial types and typologies, which were very pervasive in that era, whereas Boas did not. If Du Bois' racial typologies were flexible over the long run, and thus different from the iron cage determinism of most of his fellow citizens, Boas simply asserted that different peoples had their own histories and that there was no such thing as evolutionary racial history, as Du Bois and most

Americans (and Europeans) happened to believe. On another level, however, Du Bois and Boas agreed that there was a brighter tomorrow for their day's persons of color in American society and culture and that it was the task of the expert social scientist to show this to be the case.

And, indeed, Du Bois had a particular political program to forward, which he outlined as early as 1899 in *The Philadelphia Negro*. In Miller's acute analysis of that important book, Du Bois defined history as a process in which human action forged race-grounded social units during each race's struggle to recognize and establish its particular genius. That process helped create the drive among the European races for modernity and for creation of the United States of America as a multiracial society that included African Americans and their descendants. America was a racially diverse society, but not one in which the various units were on an equal footing. American society was hierarchical, comprising radically different and interdependent races, all capable of becoming adjusted to modern civilization through cooperation and competition within a democratic polity. That democratic polity would be held together by promises for equal opportunity for each race to realize its genius, its special ability to contribute to the commonweal. Du Bois thus thought of cities and regions as places inhabited by different races; race, not place, was central to his analysis, thus placing him thoroughly within the context of his times, in which Americans from most walks of life acted and spoke as if group identity, whether that were race or class or religion or gender or whatever other group designation might be deployed, was the key to understanding human nature and conduct, whether of individuals or of groups. Thus, the agenda he outlined in *The Philadelphia Negro* was of great portent for the future of race relations in America; social science no less than other varieties of science was informed by larger cultural constructs, which themselves acted as the lens through which he and his contemporaries, regardless of political views, interpreted the "realities" of socioeconomic and political "structure." That Du Bois spent the rest of his career campaigning for such a vision of America, mostly in futility, is well known to scholars of the black experience in America. What does matter here, as Miller reminds us, is that all historical actors, even those seen by posterity as "visionaries," are in fact locked in the time in which they have been situated and indeed change their notions of the structure of reality as these general cultural notions shift from one era to the next. And Du Bois, like most of his contemporaries, believed that the expert social scientist, like any expert in any field in which technical knowledge was the thing to be disseminated to a presumably grateful public, would identify the broad outlines of public policy, define the problems and the various remedies, and thus "inform" the masses. Social science had indeed taken its place among the traditional natural sciences of American civilization.

ZANE L. MILLER

6 Race-ism and the City: The Young Du Bois and the Role of Place in Social Theory, 1893–1901

Students of American civilization in the late nineteenth century avidly sought technical knowledge, the capacity to use science, technology, medicine, history, economics, sociology, and politics for the promotion of the general welfare and the advancement of the United States in its competition with other nations. In so doing, however, they started from a set of assumptions that depicted American society as a hierarchy of groups and parts that interacted in a mechanical way. In this mode of thought, the dysfunction of one group or part threatened the viability of the whole but without necessarily damaging the other groups or parts.

These same students shared prevailing race-ist notions that took biological races as the basic entities in the social order and arranged them hierarchically according to their "natural" superiority, mediocrity, and inferiority. In all these hierarchies the Negro race stood at the bottom of the list, a situation that made it the subject of intense scrutiny from a variety of perspectives for the purpose of determining its fitness for survival among competitive groups in a rapidly industrializing and urbanizing "modern" nation like the United States.

These students included some African Americans, such as the young W. E. B. Du Bois, who like other social scientists sought to work out a social theory that would provide a guide for social planning and the practice of social welfare in the present and the future. And like other late nineteenth-century social scientists, Du Bois gave a new emphasis to place as physical/social environment by ascribing to it a role in the processes of cultural change that in the late nineteenth century seemed a striking factor in the development of races and nations. The young Frederick Jackson Turner, for example, participated in the growing sensitivity to the role of place by arguing in 1893 that the moving American frontier had been the arena where European peoples encountered and adapted to an unfamiliar landscape and society formed by another people (the Indians). Turner argued that this recurring process of adaptation in the settlement of the continent by whites exerted an influence on European peoples so that they became the American people, a whole new nationality characterized by frontier traits that transcended enduring race-based subcultural distinctions. Those traits also persisted as the American nationality while settling the continent passed through the then conventional view of the economic stages of civilization, moving from the hunting and gathering, through the pastoral, agricultural, and mercantile into the modern industrial and (in America, thanks to the frontier) capitalistic and democratic stage.[1]

Du Bois also expressed this new sensitivity to place as physical/social environment as a factor in the development of races and nations. As a young black social scientist, however, he sought to develop a theory of the development of races and nations by focusing on the Negro race, especially the Negro race in the cities of industrial America, the place where it encountered a physical/social environment formed by another people, the Americanized European races, the most advanced of which had not only reached the industrial stage of civilization but also dominated American cities, the acknowledged centers of the industrial democratic capitalist American civilization in the late nineteenth century.[2]

Du Bois, like other turn-of-the-century social scientists, wondered whether the American Negro race had reached a stage of development that would enable it to survive in the modern city and whether in the city it could continue to develop as a race. He began this work in a series of publications on the Negro and the city between 1896 and 1901,[3] including *The Philadelphia Negro*,[4] which he wrote for the Wharton School of the University of Pennsylvania to satisfy its desire for a survey of the problems of Negroes in the Seventh Ward of Philadelphia and for suggestions for the solution of those problems, especially the problem of employment and the economic viability of the Negro in the modern industrial city.

Du Bois delivered that kind of study and more, for he surveyed all the Negroes of Philadelphia, not just those of the Seventh Ward. He also took the occasion to work out elements of his social theory about the development of the Negro race in a particular type of place in America, not on the frontier or in the rural South, but in the American city, especially in the nineteenth-century industrial democratic capitalist American city. Thus the book opens with a history of the Negro in Philadelphia, which argues that the Negro race, abducted from Africa and thrust into an advanced stage of civilization among more advanced races, had displayed the capacity to start its climb through the stages of civilization in America, only to be stopped twice in its tracks by virulent outbursts of discrimination and repression. The first occurred during the mid-nineteenth-century influx of European peasants, and the second happened during the later influx of American Negro peasants and city dwellers from the industrializing South.[5] These factors, he argued, explained the American Negro problems in cities, which he defined as the backwardness of the higher Negro classes; the sharp divisions among the Negro classes; the large number of Negroes in menial occupations; and vice, crime, and idleness in the Negro slums in the city's Seventh Ward.

Du Bois treated these problems as symptoms of the unnatural retardation of the progress of the Negro race in America through the stages of civilization toward the urban industrial and democratic period but did not systematically explicate the stage of development the American Negro race occupied in Philadelphia at the turn of the century. Nonetheless, he said enough to identify that stage as "medieval." This conception appears, for example, in his depiction of the Negro "aristocracy," a term he used not as an ironic metaphor for the race's upper class but as an authentic historical label indicating that the upper class ought to play the role of a "real" aristocracy in leading the race to the next stage of development, one he noted that all aristocracies had been slow to learn. As a result, he observed, ties of mutual interest did not hold together the Negro social classes. "Instead," we have "classes who have much to hold them apart, and only community of blood and color prejudice to bind them together. If the Negroes were by themselves [not living in a modern American-dominated city] either a strong aristocratic system or a dictatorship would for the present prevail. With, however, democracy thus prematurely thrust upon them, the first impulse of the best, the wisest and richest is to segregate themselves from the mass."[6]

The classification of the Negro as "medieval" carried a dual meaning. It suggests that Du Bois believed that the same factors that had retarded the progress of the Negro race through the stages of economic civilization had also retarded the progress of the Negro race through the stages in the development of its "genius," a genius that it, like other races, carried in its blood. Indeed, Du Bois had made that case explicitly in an essay on the conservation of races that appeared in 1897, shortly after he started work on the project that produced *The Philadelphia Negro*.

That essay advocated the conservation of the geniuses of the races. In it Du Bois argued that mankind began as several races sharply distinguished by physical traits, living a nomadic life in roaming communities of blood.

As they settled in cities these races became more similar in their external physical appearance but tended to differentiate and break into subraces according to the tendency of their geniuses. This process continued as cities came together as nations and as nations developed through the stages of civilization toward urban industrialism. By 1897, said Du Bois, the world contained eight races: the Slavs of eastern Europe, the Teutons of middle Europe, the English of Great Britain and America, the Romance "nation" of southern and western Europe, the Negroes of Africa and America, the Semitic people of western Asia and northern Africa, the Hindus of central Asia, and the Mongolians of eastern Asia. To this list he added other minor race groups, such as the American Indians, and several created by the splitting of the larger races into smaller territorial groupings, which divided the Slavs into Czech, Magyar, Pole, and Russian; the Teutons into German, Scandinavian, and Dutch; the English into the Scots, Irish, and the conglomerate American; and the Negroes into the Mulattoes and Zamboes of America and the Egyptians, Bantus, and Bushmen of Africa.

Each of the eight great races, Du Bois contended, carried and exhibited a "spiritual" continuity and cohesiveness. Each was bound together, according to the young Du Bois, first by a race identity and common blood; second by a common history, laws, religion, and similar habits of thought; and third by a "conscious striving together or certain ideals of life." These ideals represented the genius of the race, which among the advanced races (all white) manifested itself at the national stage of race development and during the "modern" industrial stage of civilization. In 1897, wrote Du Bois, the English nation stood for constitutional liberty and commercial freedom, the German nation stood for science and philosophy, and the Romance nation stood for literature and art.

The less advanced race groups (including the Slavic), Du Bois added, had not yet reached the stage of realization of their genius, but continued to strive, each in its own way, to develop for civilization "its particular message, its particular ideal." Though the complete message of the whole Negro race had not yet emerged, Du Bois argued that it could be realized by "the development of Negro genius, of Negro literature and art, of Negro spirit," which could be attained only by "Negroes bound and welded together, Negroes inspired by one vast ideal." Du Bois assigned the role of leadership in these tasks to "the advance guard of the Negro people—the 8,000,000 people of Negro blood in the United States of America," whose destiny rested not in a "servile imitation of Anglo-Saxon culture" but in a "stalwart originality which shall unswervingly follow Negro ideals."[7]

Du Bois did not and could not know what such striving might yield. But he thought the American Negro had already shown some characteristic traits and talents that provided clues to the genius of the Negro race. These included the "sweet mild melodies of the Negro slave," fairy tales of African derivation, a simple faith and reverence that stood out "in a dusty desert of dollars and smartness," and a taste for city life manifested in a tendency to move from the southern countryside to small cities of the South, thence to the larger cities of the South and thence to large northern cities, the most "modern" in America.[8]

Whatever the Negro genius might be, Du Bois thought that it could be realized in the United States without a "fatal collision" of the races and that the leadership in that process would come from city Negroes. He argued that the races inhabiting the United States, including the Negro, already shared the same laws, language, and religion, and that if a satisfactory adjustment could be made for the Negro in economic life, "two or three great national ideals" might thrive and develop in the same country. On these grounds the young Du Bois contended that the American Negro wanted neither to Africanize America nor to bleach Negro blood in the flow of white Americanism.[9] The American Negro, he asserted, wished only to be welcomed as a "co-worker in the kingdom of culture, to escape both death and isolation" and to secure in the multiracial cities of the United States the opportunity to realize its racial genius and to move into the democratic industrial capitalist stage of civilization without "being cursed and spit on."[10]

If given that opportunity, Du Bois in effect contended, the modern American city could be for the American Negro race what the American frontier had been for the white European races. In the modern American city the American Negro race had already adapted to the laws, language, and religion of the modern American nationality and had shown signs of the capacity to make a living in a democratic industrial capitalist economy. In addition, by moving to the city, the masses of the American Negro race joined the Negro aristocracy in the "medieval" stage of development of the genius of the race, the first step in the accentuation of its genius. Having taken that step, American city Negroes needed only to perfect their social organization, an achievement that would facilitate their economic progress and start the renaissance of the Negro race. That renaissance would lead to the realization of the unknown genius of the Negro race, an event that would earn for the Negro the respect of whites, improve race relations in the United States, and ultimately, as American Negroes led Negroes in other places to the renaissance, improve race relations internationally. At that far distant moment of racial equality, the Negro and white races in the United States would confront the opportunity to lead in the creation of a universal humanity encompassing the genius of all the races and unified by the striving to realize some common purpose, which Du Bois hoped would be the ideal of the brotherhood of man.

Though not fully explicated in *The Philadelphia Negro,* this theory of the development of races and nations informed the recommendations for handling the Negro problems in Philadelphia that Du Bois laid out in the final chapter of the book. There he asserted, for example, that the Negro race in America would not, as some believed, die out, for "a nation that has endured the slave-trade, slavery, reconstruction and present injustice three hundred years, and under it increased in numbers and efficiency, is not in any immediate danger of extinction." And there too he cited "voluntary or involuntary emigration" as an impractical "dream of men who forget that there are half as many Negroes in the United States as Spaniards in Spain."

Given that, Du Bois noted, "several plain propositions" seemed "axiomatic." First, "the Negro is here to stay." Second, it "is to the advantage of . . . black and white, that every Negro should make the best of himself." Third, the Negro should strive to "raise himself by every effort to the standards of modern civilization and not to lower those standards in any degree." Fourth, white people should "guard their civilization against debouchment by themselves or others but recognize that it could be done in a way not to hinder and retard the efforts of an earnest people to rise" simply for the lack of faith "in the ability of that people." Fifth, "with a spirit of self-help, mutual aid and co-operation, the two races should strive *side by side* [emphasis added] to realize the ideals of the republic and make this truly a land of equal opportunity for all men."

His vision established, Du Bois then discussed the duty of the "two races." Negroes faced "an appalling work of social reform." They had "no right to ask that the civilization and morality of the land be seriously menaced for their benefit," for "men have a right to demand that the members of a civilized community be civilized; that the fabric of human culture, so laboriously woven, be not wantonly or ignorantly destroyed." Du Bois did not expect "complete civilization in thirty or one hundred years, but at least every effort and sacrifice possible" by Negroes "toward making themselves fit members of the community within a reasonable length of time; that they may early become a source of strength and help instead of a burden."

Du Bois here, though he did not say so, was talking about Negroes becoming an interdependent yet self-sufficient racial unit, as his elaboration of the point indicates. "Modern society," he noted, "has too many problems of its own, too much proper anxiety as to its own ability to survive under its present organization, for it lightly to shoulder all the burdens of a less advanced people." It therefore "can rightly demand that as far as possible and rapidly as possible the Negro bend his energy to the solving of his own social problems—contributing to his poor, paying his share of taxes and supporting schools and public administration." In return, "the Negro has a right to demand freedom *for self-development, and no more aid from without* [emphasis added] than is really helpful for furthering that development." But "the bulk of the work of raising the Negro must be done by the Negro himself, and the greatest help for him will be not to hinder and curtail and discourage his efforts."

Then Du Bois ticked off a list of particular things Negroes should do. Negro homes must "cease to be, as they often are, breeders of idleness and extravagance and complaint" and become places where the "homely virtues of honesty, truth and chastity . . . and self-respect" are taught to a people "whose fellow-citizens half-despise them . . . as the surest road to gain the respect of others." Negroes should also cooperate to provide jobs for their sons and daughters and strive to establish "rational means of amusement" for "young folks" in homes and churches and elsewhere. He suggested in addition preventive and rescue work by Negroes, including the keeping of young and unchaperoned girls off the streets at night, warning against the lodging system, urging home ownership outside crowded and "tainted neighborhoods," lecturing on health and habits, denouncing gambling, and acquainting the mass with sewing schools, mothers' meetings, parks, and airing places. And Negroes should counsel other Negroes on expenditures, discouraging the wasting of money on dress, furniture, elaborate entertainments, costly church edifices, and "insurance" schemes, while encouraging home buying, educating children, giving simple healthful amusements to the young, and saving against hard times.

But "above all," said Du Bois, "the better classes of Negroes should recognize their duty toward the masses" and remember that the "spirit of the twentieth century is to be the turning of the high toward the lowly . . . , the recognition that in the slums of modern society lie the answers to most of our puzzling problems of organization and life, and that only as we solve these problems is our [racial] culture assured and our [racial] progress certain." Especially in the city, he added, the Negro does not appreciate this, for "his social evolution in cities like Philadelphia is approaching a medieval stage when the centrifugal forces of repulsion between social classes are becoming more powerful than those of attraction." This he could understand given the powerful past and present force of prejudice in America. Yet "the Negro must learn the lesson that other nations have learned so laboriously and imperfectly, that his better classes have their chief excuse for being in the work they may do toward lifting the rabble," especially "in a city like Philadelphia which has so distinct and creditable a Negro aristocracy."

Finally, Du Bois recommended for Negroes the cultivation of "a spirit of calm, patient persistence in their attitude toward their fellow citizens rather than of loud and intemperate complaint" and "some finesse" in advising both their fellow citizens and whites of error. It will not, he said, "improve matters to call names and impute unworthy motives to all men" and would be helpful to remember that "social reforms move slowly."

As for the duty of whites, Du Bois began by warning them once more of their preoccupation with the question "which just now is of least importance, . . . that of the social intermingling of the races." By that phrase he meant the mixing of what he saw as entities that were and for the foreseeable future and longer would and should be separate, interdependent, and self-sufficient racial units. As he put it, the old question, "Would you want your sister to marry a Nigger?" comprised a problem itself, and an unnecessary one, because it still stood "as a grave sentinel to stop much rational discussion." Since intermarriage and other kinds of social mixing were not likely to occur because virtually no one wanted or advocated them, it was now a "little less than foolish" to consider these issues, although "a century from to-day we might find ourselves seriously discussing such questions of social policy."

Much graver immediately was the issue of the "natural" repugnance of whites to close intermingling with "unfortunate ex-slaves," especially the migrating freedmen recently arrived in Philadelphia. This repugnance, claimed Du Bois, too often yielded a "discrimination that very seriously hinders them from being anything better." Objection to their ignorance, for example, he approved, but only so far as it did not create that ignorance and lead to actions "keeping them ignorant." Here Du Bois came down exclusively on discrimination in employment, the exclusion of Negroes from work, their confinement to menial tasks regardless of training, talents, and aspirations. To him it was as "wrong to make scullions of engineers as it is to make engineers of scullions," and a "paradox of the time that young men and women from some of the best Negro families in the city—families born and reared here and schooled in the best traditions of the municipality have actually had to go to the South to get work, if they wished to be aught but chambermaids and bootblacks." Such a policy led only "to crime and increased excuse for crime . . . , increased poverty and more reason to be poor, . . . increased political serfdom of the mass of blacks to the bosses and rascals who divide the spoils. . . . Surely here lies the first duty of a civilized city."

The second and final duty he prescribed for whites in their efforts for the uplifting of the Negro involved recognizing "the existence of the better class of Negro" and gaining its "active aid and co-operation by generous and polite conduct." He urged "social sympathy" between "what is best in both races." The better class of Negro wanted neither help nor pity but a "generous recognition of its difficulties." Composed of men and women "educated and in many cases cultured," this class could be a vast power in the city and the only power that could successfully cope with many phases of the Negro problems. But their aid could not be won out of "selfish motives, or kept by churlish and ungentle manners; and above all they object to being patronized."

Du Bois believed this point of sufficient importance to provide illustrations of how not to approach Negroes to gain their confidence and cooperation, how to avoid expressions of "unconscious prejudice and half-conscious actions" by those who do not "intend to wound or annoy."

One is not compelled to discuss the Negro question with every Negro one meets or tell him of a father who was connected with the Underground Railroad; one is not compelled to stare at the solitary black face in the audience as though it were not human; it is not necessary to sneer, or be unkind or boorish, if the Negroes in the room or on the street are not all the best behaved or have not the most elegant manners, it is hardly necessary to strike from the dwindling list of one's boyhood and girlhood acquaintances or school-day friends all those who happen to have Negro blood, simply because one has not the courage now to greet them on the street. The little decencies of daily intercourse can go on, the courtesies of life be exchanged even across the color line without any danger to the supremacy of the Anglo-Saxon or the social ambition of the Negro. Without doubt social differences are facts not fancies and cannot lightly be swept aside; but they hardly need to be looked upon as excuses for downright meanness and incivility.

Thus Du Bois concluded his study and his advice on the duty of whites. For them he suggested in summary a "polite and sympathetic attitude toward these striving thousands; a delicate avoidance of that which wounds and embitters . . . ; a generous granting of opportunity . . . ; a seconding of their efforts, and a desire to reward honest success." That, "added to a proper striving" by the Negro race, "will go far even in our day toward making all men, white and black, realize what the great founder of the city meant, when he named it the City of Brotherly Love."

Whatever Penn may have meant, it seems clear what Du Bois meant. He defined history as a process in which human action forged race-based social units during the struggle of each race to realize its genius. That process led to a drive for modernity among the races of Europe and to the creation in the United States of a multiracial society, one that included imported Africans and their descendants. The history of the United States, according to Du Bois, consisted of efforts, not always successful or enlightened, by race leaders to manage and mold the races into a coherent entity. This view of history also suggested to Du Bois that these efforts would yield an uncertain future in which others would manage and arrange racial relations according to their own lights.

From this perspective on race, history, and the future Du Bois sought to establish a racial policy for his generation that would guarantee both the survival, integrity, and progress of the Negro race and the coherence and progress of the United States. And he saw the United States at the turn of the century as a racially diverse society, but not as a society of equal racial units in the sense that each ranked as units equivalent in weight and presenting therefore legitimate alternatives to the others. Rather, he saw American society as comprised of a hierarchical structure of internally diverse, cohesive, self-sufficient, and interdependent (as opposed to dependent) races, each different from the other in genius but all capable of attaining an acceptable level of modern civilization through cooperation and competition within a democratic polity. Such a diverse entity would be held together by the guarantee of equal opportunity to all races in the competition for a place in the modernizing hierarchy of races and by an appropriate distribution of respect for those races on the lower rungs of the hierarchy of modernism, races that might or might not catch up to the races at the top of the ladder of civilization, but that in any case would ultimately realize their genius. Negroes in Philadelphia, like others elsewhere in America, Du Bois contended in the 1890s, needed to be civilized, but it could be done, and the crisis of the American Negro and the United States could be averted if each race took care of itself while giving the others a clear field and a fair chance.

In this perspective it seems appropriate to classify the young Du Bois as a characteristic turn-of-the-century student of race, city, and regional life. That is, he conceived of cities and regions as physical and social environments inhabited by interdependent agglomerations of races, each of them stamped by distinctive physiological, psychological, and social characteristics derived from their biological and historical heritage and each carrying within it a race-based genius. While he doubted the availability of the essences of these geniuses for direct study and analysis, he believed in the possibility of identifying and encouraging them by monitoring statistically the behavior of the races in the present and future and by the analysis of their heritages and tendencies through the study of their pasts, a project he proposed for the study of the Negro race in America and that he sought to carry out at Atlanta University after completing *The Philadelphia Negro*.[11] Such a study of a race and its relationship to other races would yield, Du Bois thought, important scientific information and conclusions about the race and the cities and regions it inhabited. It is important to remember, however, the primacy attributed by the young Du Bois to race and to the "givenness" of the character and genius of the race. This emphasis meant that the condition of a particular city or region depended on its mixture of races and their interrelationships. Those interrelationships depended to a large extent upon the stage of each race in civilization, the degree to which each race had realized its own genius, and the degree to which the advanced races could understand the necessity for tolerating the development of the civilization and genius of the backward races. This kind of racial tolerance ranked for Du Bois as the necessary precursor to the emergence ultimately of a universal community of purpose dedicated to the brotherhood of man, "that perfection of human life, for which we all long, that one far off Divine event." [12]

In *The Philadelphia Negro,* however, Du Bois looked to the shorter run. He depicted Negroes as a race-based social entity containing the potential and adaptability necessary for successful participation in the competition of modern life in a modernizing nation. He projected, that is, a strategy of equal opportunity among competing races—by which American Negroes might become and function as a separate, internally integrated, cohesive, and interdependent social unit somewhere along the lower ranks of the hierarchy of races inhabiting the United States. But his immediate concern focused on a crisis, a fear that the direction of racial policy within the United States might fall irrevocably into the hands of those who would deny this opportunity to Negroes and thereby destroy the potential for the civilizing of the Negro race and the development of its genius in America. Such an outcome would also disrupt the potential for peaceful development of a homogeneous community of purpose in the United States, a process, Du Bois contended, that might especially benefit from the preservation and utilization of some of the products of the development and conservation of the genius of the Negro race.

Despite the preoccupation of the young Du Bois with the condition of the Negro race in American cities, it seems inappropriate to think of him as an "urban" historian or sociologist, for he seems not to have thought of himself as an urban historian or sociologist or as an urban anything. Indeed, no one then thought in those terms. That is, the concept of things urban as we usually think of it emerged in the 1920s, when culture became embedded in place, so that people, place, and culture became inseparable. In the second quarter of the twentieth century, where you came from rather than your race determined who you were. In this context of place-based cultural determinism, social theorists and planners contended that urbanites, regardless of their race or nationality, differed culturally from rural dwellers, just as they contended that Southerners differed culturally from Westerners. In that same context the more malleable and new concept of place-based ethnicity displaced that of race-based groups, and American society came to be seen as one that was, was becoming, or ought to become culturally pluralistic. Neither the young Du Bois nor other social theorists at the turn of the century thought like that. For them, culture stemmed from race, not place, although place as physical and social environment might condition a race-based culture by advancing or retarding its progress through the stages of civilization and/or by advancing or retarding the realization of the genius of the race.[13]

Nor did the concept of race lurk merely in the background of *The Philadelphia Negro*.[14] The young Du Bois was captivated by the idea of race and racial history, specifically by the history of the Negro race in America and especially by the experience of freedmen and their children since the eighteenth century. For Du Bois the Negro in America was not African, southern, urban, or rural in essence, but a distinct race, one shaped by its biologic legacy and its history of repression. That dual inheritance stamped it as inferior to the white races, inferior in its preparation for life in the United States as American society modernized during the nineteenth century. In other words, the American Negro race carried that dual legacy wherever it might be found, and at the turn of the century Du Bois focused his attention on the condition of that portion of the race living in American cities, especially but not exclusively on the children of those emancipated by the Civil War, who by virtue of their emancipation became a nomadic class, the most retarded of all the Negro classes but a class which by 1900 was becoming numerically dominant even among northern Negro city dwellers.

From this it follows that Du Bois' turn of the century work does not represent the precocious product of a visionary ahead of his times, part of the advance guard of "cultural pluralism,"[15] which comprised a place-based determinism that did not emerge until the 1920s. From this it also follows that *The Philadelphia Negro* was not the first of a series of works on the American Negro by great sociologists culminating with St. Clair Drake and Horace R. Cayton's *Black Metropolis* (1945).[16] That study sought to examine one of Chicago's many place-based cultures, the culture of the black ghetto, a culture indigenous to a place that stamped its residents, who remained a part of the ghetto even when they left it to take up the arduous and risky task of adjusting to the culture of another place.

Instead, the young Du Bois, like other social theorists of his day, was a race-ist, one who took the category of race as a fundamental and deterministic element in the taxonomy of social reality and who participated in the process of examining the influence of place as physical and social environment on the progress of races as they realized the geniuses within their blood and moved through the stages of civilization.[17] And he sought not equality in the near future for the Negro race in America but a fair chance for that race to prove that it could compete in a modernizing nation, become modern itself and therefore a legitimate part of democratic industrial capitalist America.[18] If given that chance, the young Du Bois concluded, the American city Negro would lead in the development of the Negro's unrealized racial genius as a contribution to the racially pluralistic society of the United States, a society that might then take the lead in forging a transnational and homogeneous community of purpose. Then, and only then, after the realization of the genius of the Negro race, said the young Du Bois, would the question of racial intermarriage become a serious question of public policy.

Introduction to Chapter 7

That the science of nutrition might ever be allied with conceptions of human racial characteristics at first seems a peculiar suggestion. Yet in his essay on the American science of racial nutrition, Hamilton Cravens makes precisely that point. Larger cultural notions—in this instance, ideas of racial differences and of the centrality of racial traits to human behavior—shaped and informed technical knowledge about human nutrition in both Europe and America. In that sense Cravens's essay follows from and corroborates Miller's on Du Bois, showing how pervasive were these notions of group identity for the individual.

Central to the events Cravens describes was the campaign that several influential individuals, including the scientists Ellen H. Richards, Mary Hinman Abel, and Wilbur O. Atwater, and the Boston businessman Edward Atkinson, conducted to have the U.S. Department of Agriculture's Office of Experiment Stations support a large number of scientific investigations of the typical diets of the many distinct races in the American national population so as to determine what the ideal racial diet for the nation's denizens should be. From one vantage point, each participant seemed to join hands with the others for highly particularistic, if not downright bizarre, and individuated reasons. These included driving down the cost of labor for the put-upon American capitalist (Atkinson), applying the laws of thermodynamics to the science of nutrition and boosting the status and power of analytical chemists (Atwater), advancing the cause of female scientists in their struggles to be recognized by male scientists (Richards), and quieting social unrest and assimilating immigrants through an Americanized diet (Abel). Obviously generalized and even tacit assumptions in the culture about how society, polity, and economy did and should work had much to do with promoting certain kinds of scientific investigations at the expense of others; or, put more starkly, some kinds of scientific work were simply inconceivable, whereas others "made sense" in that age. The resulting campaign, which stretched from the early 1890s to the middle 1900s, had different outcomes for each of the interested parties. Whether some "won" and others "lost" matters less in this analysis, finally, than the more rigorously historicist point that all were comfortably nestled in the age in which they lived.

Yet there is more to the story Cravens narrates than this. In fine he draws attention to four groups of would-be nutritionists, one in Germany and two in the United States who were chemists, and a fourth in the U.S. group whose leader vigorously criticized the chemists for not understanding the biological aspects of human nutrition but who nevertheless did not alter the fundamental notion of nutritionists of that age, that there was a normal or standard diet for each population or group or race of people or, put another way, that diet was a matter of one's group identity. In an era in which practitioners of the human sciences acted and spoke as if there was no such thing as an individual, but only groups whose members shared particular traits in common and varied among themselves around a particular distribution or mean of traits, this kind of "racial" or group-oriented nutritional science made all kinds of sense to all manner of people, whether they were scientists or not. Thus even though these groups of scientists seemed deeply involved in technical knowledge, that technical knowledge itself was informed by larger cultural constructs and notions, and here more specifically by constructs that were embraced in two distinct western national cultures. To be sure, there were important differences between the Wilhelmian Reich on the one hand and William McKinley's America on the other. Yet in each there were fairly rigid hierarchies of race, class, and religion, as well as particular notions that each race or class represented a bundle of traits and patterns of behavior. Furthermore, most Germans and Americans believed and acted as if centralization, vertical integration, efficiency, and standardization were the keys to a dynamic economy and nation state. Thus with only slight modifications here and there, a bundle of technical ideas could resonate in two national cultures within a common European and American civilization. That in turn suggests how the history of science—or of any specific body of knowledge or any cultural discourse—can be used to assess questions in comparative history; if different national cultures in a common larger civilization have parallel constructs of technical knowledge, then this suggests that the methods of the cultural historian may have a higher yield in comparative history than do those of the social historian.

The German-American Science of Racial Nutrition, 1870–1920

7

In May 1893, Ellen H. Richards, a prominent chemist at Massachusetts Institute of Technology (MIT), excitedly told her protégé Mary Hinman Abel that the "new Sec. [sic] of Agriculture has just written to me in regard to establishing food laboratories in connection with the Agric. [sic] Exp [sic] Stations of the US, [and] the whole subject is ready for a boom. We must be in it."[1] And so they were to be, in what was to become the American science of racial nutrition, according to which each and every race had its own distinctive diet that made it efficient and effective in life. Here, then, was a most interesting example of technical knowledge—knowledge about food—that was informed in its larger contours by notions of race, presumably products of the larger culture itself.

Richards was one of the first U.S. women chemists professionally employed as such. After getting what scientific education she as a woman was allowed at Vassar and MIT, she became an expert chemical analyst for New England industrial and chemical firms. In 1884 she acquired her institutional base as instructor in sanitary chemistry and engineering at MIT. Connected by business and social ties to MIT, she soon came to know Edward Atkinson, a prominent Boston industrialist and conservative Democrat. He introduced her to many potential industrial and governmental clients for her expert services. And he had the requisite connections in national politics and within the Democratic Party to launch a campaign for racial nutrition.[2]

By the spring of 1893, Richards, Abel, and Atkinson had acquired something of a public reputation as advocates of the New England Kitchen (NEK) in Boston. In 1888 Richards met Abel, and thus the problem of human nutrition, by chance—she was a member of the Lomb prize committee for the American Public Health Association that year; the Lomb prize was for the best essay in public health and how to improve it. She read Abel's essay and made sure that it won the prize. It was published as *Practical Sanitary and Economic Cooking Adapted to Persons of Moderate and Small Means* (1890). Abel and her husband, John J. Abel, a prominent pharmacologist, had lived in Europe for several years in the 1880s, especially in Berlin and Munich. There she had learned of the famous *Volkskuechen,* or eleemosynary soup kitchens for the poor. By providing cheap, nutritious, and wholesome food, these kitchens seemed to address potentially explosive social problems, at least from a middle class or bourgeois perspective, and gave her some ideas for improved public health and welfare in America. In her essay, Abel insisted that America's poor could have a better—that is, more nutritious and less expensive—diet. Abel included sample weekly menus and suggested such alternatives for expensive meats as cheese, legumes, and inexpensive cuts of meat.[3]

Richards and Abel soon became close allies. They put their ideas into action. They believed that one could prepare a separate cuisine for each social class, a sort of capitalist credo for nutrition, from each according to his ability to pay to each according to his station in life. By eating less expensive foods, like their European peers, American workers would therefore live on their wages and be more content, less tempted to support social upheaval. If the *Volkskuechen* were an inspiration for their plans, they still modified the Continental example. Their innovation, the NEK, was no soup kitchen for the poor, nor did they assume that the poor were a permanent feature of society. Instead Abel and Richards wanted an institution that would upgrade and uplift immigrants and persons of color to be "true Americans"—and truly middle class, as well. They designed it as a weapon of Anglo-Saxon middle class cultural uplift—and imperialism.

Now Edward Atkinson entered the picture. A loyal Cleveland Democrat, he was profoundly laissez-faire and anti-tariff. With some hysteria, he believed that international competition meant workers' wages could not rise, and, therefore, their abysmal living and working conditions would lead to social upheaval and revolution. Various experts told him that whereas European workers spent only about half their income on food, American laborers spent upwards of ninety percent this way. Obviously, he concluded, American capitalism could compete only if American workers would eat cheap foods as their European counterparts did.[4]

In the *Century Magazine* in 1886–87 Atkinson called for dietary reform among U.S. workers as the means of saving American capitalism.[5] He invented the Aladdin Slow Oven for his campaign. It was a box of fire-resistant materials into which a kerosene lantern was placed; foods could be cooked in it, thus saving considerable monies on fuel costs. And it was the ideal appliance because workers could prepare their inexpensive meals—their soups and stews and baked goods, that is—in it. Atkinson turned to his longtime friend Richards for appropriate recipes for the oven (and the campaign). Naturally she recruited Abel. Atkinson raised money from Boston's wealthy elite for the planned NEK; soon there was a food laboratory in which Richards, Abel, and their assistants could devise working-class recipes, all with a Yankee or all-American twist. Richards and Abel modeled the New England Kitchen after the agricultural experiment station (MIT was a land-grant institution, after all) in which experts showed a presumably grateful populace how expert knowledge could bring average folk a better life. Because Atkinson's oven took ninety minutes to reach two hundred degrees Fahrenheit, the recipes that Richards and her minions devised produced slow-cooked meals. And the meals were also bland, boring Yankee soups, stews, and casseroles with all of the life slow-cooked out of them.[6]

The NEK was not a rousing success. From 1890 to 1894, the kitchen's approximate life span, it existed chiefly in Boston but also in a prettified version at the Chicago World's Fair in 1893. Boston's masses, presented over and over again with examples of how to cook these allegedly yummy delicacies, resisted them massively as yet another example of Yankee cultural imperialism. Its cuisine assumed that there was one dietary standard that would feed all the immigrants and nonwhites. Richards believed, as did Abel and Atkinson, that the urban masses would be uplifted to an American norm or standard if they were to eat this food. In particular did the New England Kitchen differ from the *Volkskeuchen,* which Richards thought were communal or socialistic enterprises totally inappropriate for what she dubbed "the free American." More than a mere transfer of food to families from the NEK was involved. Richards, Abel, and Atkinson believed that the food would nurture American habits, customs, and notions within the immigrants and other slum dwellers by teaching them the "right" (less expensive and more nutritious) foods to prepare and consume. All meals were fixed in full public view; thus the NEK worked like an agricultural experiment station, showing the masses the right techniques. The poor would learn how to eat cheaply. They would become thrifty, hardworking, true Americans. Clearly the champions of an American science of racial nutrition wanted to create a unified American race through a uniform diet, although it was unclear whether it was the food itself, or just the ways of preparing and consuming it, that mattered more to them.

Yet the poor ate what they wanted to eat. There were kitchens established in Boston, Chicago, and Philadelphia. None were really self-sustaining. Thus those in Boston for African Americans and for white immigrants, for example, were never self-sustaining. Although one survived at Jane Addams' Hull House settlement in Chicago for about thirty years, it stayed afloat because another eating establishment in Hull House paid its bills. Even the cooking school that NEK graduates set up in Philadelphia flopped dismally.

Even before the NEK's demise, Abel, Richards, and Atkinson found a better opportunity for reforming the American diet and saving American capitalism: Grover Cleveland's return to the White House in 1893. Atkinson's credentials as a wealthy, conservative Democrat gave him rapport with the new administration, especially with Agriculture Secretary J. Sterling Morton of Nebraska, also a conservative Democrat, who believed with Atkinson that farming was a scientific venture. Lately the federal government had seemed a promising sponsor for scientific agriculture, as in, for example, passage of the Hatch Act in 1887, which gave each state and territory $15,000 per year to fund an agricultural experiment station; in the creation of the Office of Experiment Stations in 1888; and, in 1889, the Agriculture Department's elevation to cabinet rank.[7]

Atkinson and Richards allied themselves with that most relentless champion of scientific agriculture, Wilbur O. Atwater of Wesleyan University. He seemed the answer to their dreams. A key player in the recent battles over the Hatch Act and related matters, Atwater was probably the country's most distinguished nutritionist. After taking his Ph.D. in chemistry at Yale in 1869, Atwater studied at the University of Berlin with Carl von Voit, a leading nutritional chemist, until 1871. With his students von Voit explored the insight that in food, proteins differed from carbohydrates and fats because they stimulated the body's metabolic processes as measured by the rate of oxidation. Thus food was to be regarded as fuel, and for a "machine" that exemplified the laws of energetics. The chemical composition of distinct foodstuffs—the amounts of protein, carbohydrates, and fats they contained, that is—was the central issue in late nineteenth-century nutritional science. Thus the most successful organism, human or animal, was the one that got the most calories to power the machine and also got enough protein to repair worn tissue. The science of nutrition was concerned with efficiency and energetics. Von Voit's methods involved analysis of the chemical composition of foodstuffs so as to calculate the digestion and use of proteins, carbohydrates, and fats. Foods and wastes were carefully measured. In 1864 the Association of Official German Agricultural Chemists adopted von Voit's methods for the field, which insured their acceptance worldwide for decades to come.

It was Max Rubner, one of von Voit's students, who extended the first law of thermodynamics to the animal body. Rubner had become professor at Munich and was a rival of von Voit's in ways that had nothing to do with the large differences between Berliners and Bavarians. He believed that the body's metabolic processes were fueled by burning calories—an insight of the nascent science of energetics. And Rubner's ideas had a large ideological import. Rubner thus challenged the vitalists, who insisted that there was a qualitative difference between life forms and nonliving forms in nature. By arguing that the animal body behaved according to principles that also explained the working of a steam engine, or other forms of work, late nineteenth-century chemical nutritionists were winning a long war with theologians, clerics, poets, philosophers, and others concerning the nature of life and the nature of nature. They helped pave the way for late nineteenth- and early twentieth-century scientific reductionism, determinism, and positivism. Here again larger cultural notions helped to shape technical knowledge, or, more precisely, such notions helped investigators interpret the natural phenomena they observed.

This was quite evident in von Voit's *The Physiology of General Metabolism and Nutrition* (1881), in which he first discussed the relations of physiology and metabolism to how the animal machine worked and then devoted the latter forty percent of his pages to his analyses of various foodstuffs and beverages as nutritional fuels. It was even clearer (and infinitely more precise and mathematical) in Rubner's important monograph, *The Laws of Energy Consumption in Nutrition* (1902), a summary of much of his work at Munich, in which he spelled out the specifics and the implications of thermodynamics as applied to the animal (and human) body. Rubner's coining of the isodynamic law, his measuring of the exact caloric equivalent per gram of each foodstuff, and his measuring of the specific dynamic effect of foodstuffs—all these technical accomplishments, that is—nevertheless fed his pride in having undercut vitalism. And such a framework—that there were machines, animal and human, that worked according to certain general principles of heat and energy—made it easy to believe that there were distinct races that nevertheless followed particular patterns or natural laws, perhaps the underlying assumption of contemporary western anthropology—and colonialism too. For that matter, the notion that efficient production was a positive good within the western bourgeoisie or middle class in the nineteenth century was but another example of a notion from the larger culture helping to shape and inform technical knowledge.[8]

Upon returning to the United States in the early 1870s, Atwater launched his career as perhaps America's most ambitious proponent of agricultural science. In 1873 he became professor of chemistry at his alma mater, Wesleyan University, and two years later joined the Connecticut Agricultural Experiment Station Board of Control, a position he held until 1905. He was an able scientific entrepreneur. Soon he was conducting research for the United States Fish Commission to promote fish culture, which meant increasing the supply of fish and promoting the New England fishing industry. He approached the problem from two interrelated angles. First, fish were a source of superphosphates, which since mid-century had been used as agricultural fertilizers—a topic of obvious interest to Atwater. Second, fish were an inexpensive source of protein, and, following the now conventional wisdom of his science, protein was a source of energy for muscle tissue, helping to rebuild and nourish it, thus making possible better health, strength, and productivity.

An outcome of Atwater's work and that of his associates was the discovery that nitrogen helped promote the growth of proteins, which he confirmed in 1881 with legumes, arguing (correctly) that they were a plentiful source of protein. Like his friend and Berlin classmate Rubner, Atwater was chiefly interested in applying the laws of contemporary physical science to human nutrition. With that typical late nineteenth-century enthusiasm for reducing problems of organic nature and sentient beings to principles of inanimate nature and mechanistic laws, he was sure that he could apply the laws of thermodynamics to human nutrition, thus following contemporary scientific fashion. To Atwater any sentient being, physiologically or metabolically speaking, was simply an engine that required fuel. To continue the analogy, food was the fuel that sentient beings consumed to produce the energy that enabled them to be active organisms. The problem of human nutrition, for Atwater and all nutritionists was how to obtain the least expensive and most nourishing diet for mankind.[9]

In 1886 Atwater made an important trip to Europe to confer with scientific colleagues about the problems of human nutrition. He spent much of his time in Munich with Rubner at the university's physiological laboratories, perhaps then the world's most celebrated facility in the field. Atwater learned firsthand the most up-to-date work on human nutrition available, including a wonderful new gadget for researches, the world's first successful calorimeter, which Rubner had designed. It was essentially a room within a room filled with marvelous instruments for measuring the intake and outgo of gases in the sentient beings contained therein so that their consumption of energy could be measured. Atwater wanted one for his laboratory. No less important was what Atwater later called the "dietaries," or records of the precise consumption of the volume of foods and of their chemical analysis, by different classes, nationalities, and races in Europe.

Atwater extolled the new science's virtues in the *Century* in a series of articles hard on the heels of Atkinson's articles there. He presented nutrition as a European and more specifically a German artifact; he knew such disinformation would give his arguments more legitimacy with his opponents and with the educated public, who were still used to thinking of European culture as superior to American culture. He made several arguments—now people could begin to understand the science of nutrition; in particular they could apply it to pressing social problems. One could come to grips with the food economy of mankind. And, he argued, the food economy thus revealed demonstrated that throughout Europe members of the various social orders and classes and nations ate less expensive food than their social counterparts in the United States.

Atwater charged that American workingmen were a wasteful lot. They could readily find cheaper foods and save their money for life's other necessities. They should eat far more protein, the source of tissue regeneration, than carbohydrates, the source of mere fuel, and they could eat infinitely cheaper proteins, such as legumes or cheese, just as did their European counterparts. In sum, he insisted, the nutritional differences between expensive and cheap foods disappeared under chemical analysis in the laboratory; the body absorbed and used fats, carbohydrates, and most proteins in about the same way. One should eat only as many of the necessary nutrients as were needed to perform their specific physiological tasks. Anything else was wasteful. Atwater excoriated Americans for eating more carbohydrates than necessary, especially as sweets, and, he continued, the poor ate too little protein. Americans of all classes and races, by wise economies and with expert knowledge of nutrition, could spend less on food and be better nourished. Above all, Atwater urged that people should select food that was good for them, regardless of its taste or appeal. Food's chemical composition, and its relation to the food economy, were what mattered.[10]

In the fall of 1893, then, the campaign for racial human nutrition got under way. Secretary Morton had always supported the notion of science as the key to better and more productive agriculture. In this he and Atwater were natural allies. Atwater had faithfully represented the college professors of science, or, more precisely, agricultural science, in battles over agricultural institutions. Those who opposed the professors were agricultural leaders who believed that farming would benefit most by being made systematic and efficient, those important shibboleths of late nineteenth- and early twentieth-century American culture and society.

Indeed, this conflict between the champions of scientific agriculture and those espousing agriculture-as-a-business was the touchstone to the age's politics of agriculture. Morton once referred to the champions of systematic agriculture as charlatans who "farmed the Farmer." Morton was concerned about waste and abuse in the department, especially its free seed program for favored farmers and its apparent spoils politics. Once in office he eliminated these practices, thus freeing more funds for scientific agriculture.

Given the contentiousness of agricultural politics then, there was still controversy over alleged waste in the agricultural experiment stations; Morton, for one, believed that the farming-as-a-system types in some state experiment stations deliberately misused Hatch Act funds, and his opponents believed that of him and his allies. Thus a program to shape up the system of stations, as well as to improve agriculture through science and to save American capitalism to boot, appealed to Morton and his allies in the administration and the Congress. Obviously such a program had to have all of science's earmarks. It could not merely be businesslike, systematic, and efficient. Morton was thus quite receptive to such proposals as Atwater and Richards, through Atkinson, would make.[11]

In September 1893, in the Office of Experiment Stations' *Bulletin,* Atkinson published his call for a new science of racial nutrition. Atkinson's program, which was also that of Atwater, Richards, and Abel, assumed that humankind was a larger whole of distinct races, each with their unique assets—and liabilities. "Each race . . . through a process of natural selection, appears to have reached a unit of food, simple or compound, in which the 'nutrients', so-called, are to be found in about the right proportions," he wrote. Each of these food units had been produced or combined in such a way as to assure the maximum of nutrition at the least cost, and each national food that he enumerated had the principal nutrients of nitrogen, starch, and fat in substantially the right proportions, to make the typical members of those races or nationalities the most efficient they could be, he insisted.[12]

Atkinson enumerated these national or racial diets. Thus the English had a national ration of wheat bread, with cheese rather than butter, whereas the Scots had oatmeal, milk, and salt. The French, on the one hand, had as their optimal national diet soups and stews containing heavy portions of legumes such as beans or peas, but the Italians had macaroni made from wheat and polenta from corn, cooked with cheese made with skim milk. In Asia, the Indians and the Chinese all had rice combined with beans or peas, but the Japanese performed best with miso soup made with fermented barley, and with legumes. In North America the Canadians did best with porridge made with dry peas, coarse wheat crackers, savory herbs, and a little pork, whereas the New Englanders prospered on a racially appropriate diet of baked beans and brown bread, or codfish balls, salt pork, and potatoes. And, finally, insofar as the persons of color were concerned, for New Orleans Creoles the optimal diet was red beans and rice, whereas blacks in the southern United States required, for optimal performance, bacon, corn meal, and molasses. Although doubtless Atkinson, the public man, got these racial diets from Atwater—and probably Richards and Abel too, the scientists or experts in technical knowledge, that is—their intellectual origins and sources were less in science than in the larger culture. They gave the appearance of well-known racial or ethnic stereotypes in very skimpy scientific dress indeed.

Atkinson had a program. It involved figuring out what was the best American diet, "best" meaning the most efficient and the most cost-effective—setting a norm for all, in other words. The work of Atwater and others on human and animal nutrition, he charged, showed that Americans wasted their food and spent too much money on it. Our national dietary was very one-sided, he insisted, and the farmer and the consumer lost heavily as a result. We produced too much fat, starch, and sugar and too little protein. Ultimately we needed to find out what was the best diet for the American race or, more precisely, for the races that exist in the American population. A much more thorough and systematic investigation should be launched to collate and bring together information scattered here and there, and new work was needed on cooking inexpensive foods as well, as the New England Kitchens demonstrated. The project should involve the cooperation of many investigators. Hence he recommended a large and aggressive cooperation program in human nutrition, to be coordinated through the Department of Agriculture's Office of Experiment Stations, which Atwater controlled indirectly. There were at least three general questions for each race in the American population. First, what were the best means of producing the highest amounts of human energy with the least waste? Second, was a variety of food better for muscular or brain work? And, third, what was the relation of exercise (aside from work) to human nutrition? Congress, he argued, should allow the agricultural experiment stations to do experimental work on human nutrition. This had already been done in Europe, complete with government sponsorship, with an eye to economical food consumption; the problem there was scarcity. Our problem was waste. "With respect to 90 percent of the population of this country it has been conclusively proved that from 40 to 60 per cent of the average income of each family is expended on the purchase of food materials," Atkinson concluded.[13]

Congress passed the program. It appropriated funds, starting with $10,000 for a respiration calorimeter for Atwater at Wesleyan, to be run through the Office of Experiment Stations.[14] Atwater had left the Office of Experiment Stations in 1891 to tend to matters at Wesleyan, making sure that allies still ran things in Washington, but he was special editor for foreign work, which meant European (and German) nutritional science. Thus Atwater ran the program from every angle. Through the Office of Experiment Stations he had the European work gathered and collated. Then his protégés at the various state experiment stations were to do the many dietaries, or daily diets, of the multiple races and classes of the American national population. The law enacted stipulated that the secretary of agriculture, using the system of experiment stations throughout the country and the Office of Experiment Stations, should report on the nutritive value of the various foods used for human consumption, "with special suggestion of full, wholesome, and edible rations less wasteful and more economical than those in common use."[15]

The human nutrition campaign, so launched, solved a number of Atwater's political problems as a Tsar of American science. A formidable problem was Ellen Richards herself, who was a professional rival in New England—a Tsarina of American science, so to speak. She disdained grubwork in the laboratory. She liked running things and founding movements that she could control. Above all, she liked being in the public eye, as her letter to Abel in 1893 suggested. She had invaded Atwater's turf, and he did need her, especially to get information on recipes and grubworkers in the laboratory. But he outfinessed her by getting Atkinson to enact his program and having the Office of Experiment Stations and the state experiment stations sponsor the human nutrition campaign. These were neither political processes nor public institutions Richards could influence in the slightest, let alone control. While Atkinson recommended Richards' New England Kitchens in his 1893 pamphlet, they were dying institutions and everyone involved knew it. Richards did participate in the human nutrition campaign by training bright young women in her nutrition laboratory to work alongside Atwater's male protégés in the agricultural experiment stations, but even after Atwater withdrew in 1904, it was obvious that her role was distinctly secondary. Ultimately Richards organized the sexually segregated (and thus safer) field of home economics at the Lake Placid, New York, conferences in the early 1900s. No longer did she have to serve ideas to men without attribution, let alone fanfare, or stay in the boring laboratory. She could blossom as a public advocate of more opportunities for women and be in the limelight. She loved it. She died in 1911, much revered as the founder of home economics, a feminized parallel to systematic farming (and progressive cultural uplift as well) if there ever was one. Mary Hinman Abel, her constant companion and memorialist, lived on well into the 1920s writing the same kinds of popular treatises about family nutrition as she always had and singing the praises of Richards as the orchestrator of the American home economics movement.[16]

And, when Congress lavished so much money on Atwater's program through the experiment stations—first $15,000 and then $20,000 per year—Atwater appointed his own protégés there and crowded out his opponents, the devotees of farming-as-a-business. In 1906 Congress passed the Adams Act. It increased the appropriations for the experiment stations and permitted only natural-science kinds of work at them, thus cutting out all of the remaining farming-as-a-business personnel at the stations, which ended the decades-long war between the champions of scientific agriculture and those favoring farming-as-a-business. For advocates of farming-as-a-business, their professional descendants in the agricultural colleges became agricultural economists (and other agricultural social scientists) who returned to the experiment station trough in the decades to come. Atkinson soon found other interests. He retired from the field in 1904 and died in 1907 of stomach cancer.[17]

To 1907, Atwater and his associates carried on more than one hundred distinct research projects resulting in fifty-six different publications appearing until the mid-1910s in the *Bulletin* of the Office of Experiment Stations. Atwater hoped that once all the dietaries of the various classes and races of the U.S. population had been studied, a national diet or series of national diets could be drawn up for maximum efficiency and economy—those two bywords of the era spanning the 1870s to the 1920s—for the nation.

Thus in 1895 Atwater himself sketched the research program's general outline. Food was the chief item of expense of every household, he argued, even though people understood very little of what their food contained and what it did to or for them. The net result was "great waste in the purchase and use of food, loss of money, and injury to health." He insisted that it was necessary to know how much of each of the ingredients of food was needed to supply the demands of people of different age, sex, and occupation and how to adjust the diet to the wants of the user. Disease could result from overeating or other dietary fallacies. The body conformed to the laws of conservation of energy; intake always equaled outgo. The quantity of heat set free when a given food material was burned with oxygen, he insisted, might be taken as the measure of the fuel value of that food material; this could be measured with the calorimeter, whether at Munich or Wesleyan.

Therefore, there were two chief angles to the campaign for racial nutrition. At Wesleyan, Atwater worked with his calorimeter. In 1897 he and a colleague published a very systematic and comprehensive survey of the scientific literature on the approximately 3,600 experiments in nutritional physiology in Europe and the United States in which the balance of one or more of the factors of income and outgo was calculated. Although they could not come to any definite conclusions as yet, Atwater and Langworthy nevertheless remained cheerfully optimistic that it was only a matter of time until this great puzzle of nutritional and human science would be resolved.[18] In 1899 Atwater reported that as his earlier experiments had measured animal and human income and outgo of food matter, now he was measuring intake and outgo of energy, thus providing a "gradual approach toward a demonstration of the application of the law of conservation of energy in the living organism."[19] So much for what Atwater clearly regarded as the "basic" or "theoretical" side of the human nutrition campaign, his campaign against vitalism in science and philosophy. Ever the eager politician of science, he also managed, from Wesleyan and through his connections at the Office of Experiment Stations in Washington, the "practical" or "applied" work at the various state agricultural experiment stations.

It was Atwater's former students and other allies at the various experiment stations throughout the nation who conducted what he called the "dietaries," thus carrying out the program's second part. These were social investigations of the food that a "typical" family representing a particular socioeconomic class or a race would consume, together with recommendations on what the family should eat to get the greatest nutritive value for their money. Atwater thought that any particular group affiliation or identity might be important; as many different kinds as existed had to be tested to verify or modify his hypothesis. Race, class, gender—these terms were real to him—but so was virtually any other category that one could recognize. Clearly he was impressed by the dietaries he had seen at von Voit's laboratory in Berlin; of the total of 491 that von Voit and Rubner had collected from the United States, Europe, and Japan, he had scrutinized carefully some 338. His analyses then had convinced him that group identity, including race and nationality, mattered in terms of diet.[20]

The methods and procedures in the dietary studies were simple. Experiment station scientists identified subjects, recorded the amount of food each person in the group consumed for about ten days, did chemical analysis on foodstuffs identical to those consumed, and compared the results against standards of consumption that Atwater had worked out. Atwater had calculated that, in descending quantities, one meal of an adult woman would be eight-tenths of the meal of a man at moderate work, a child's meal six-tenths of that man's meal, and so forth, to infants with three-tenths of a share, and men doing no work would eat less than men doing heavy work. Intake and outgo were thus measured, and there were objective standards, so Atwater believed, by which to measure each race or group.[21]

First there were the dietary studies of white Americans of varying social ranks. At the Maine State College and Agricultural Experiment Station, in Orono, Whitman H. Jordan, the station's director and specialist in animal nutrition, did a complicated study for his old chief, Atwater, of cheap and dear protein, among a particular class: college students. When the project began, Jordan weighed all food and food materials in the boarding house; as more food was purchased or wasted, it was weighed as well, and at the project's close, another inventory of all food materials was taken. Jordan controlled the source and supply of animal food (and protein) and compared high- and low-cost foods as a source of protein, paying particular attention to the influence of the free use of milk as a low-cost animal food. The project took 209 days, and the number of meals studied was 14,745, or forty years of meals for one man. There were five successive stages of diet, including ordinary conditions, high- and low-priced foods, and unlimited and limited use of milk. His general conclusion was that milk was an economical and nutritious source of animal protein, clearly superior for less advantaged families than other possibilities. He also found that about 69 percent of the cost of food was in animal foods; hence the use of milk was crucial to lowering the cost of foods.[22]

Nor was this all. Winthrop E. Stone, a chemist turned vice president at Purdue University, compiled a comparative dietary of a professor at the university with that of a mechanic's family. He concluded that in both cases there was an insufficiency of protein in the diet (one of Atwater's major points), although the professional man ate more protein than did the mechanic and his family. In Atwater's comment on Stone's study, he noted that Stone's findings were congruent with other, similar studies and that dietaries of other professional men showed that they ate slightly less food than did mechanics. At Rutgers University the director of the New Jersey Agricultural Experiment Station, Edward B. Voorhees, who had been an assistant with Atwater at Wesleyan, had a dietary study of a mechanic's family in New Jersey executed. He argued that the family consumed about what other mechanics' families ate but less than what professional men's families did, thus underlining, as had Stone, the importance of disposable income for food purchases.[23] At the University of Missouri Sidney Calvert and two students ran two projects on nutrition of university students; in one, they queried several hundred about their bread and meat consumption and received 258 replies. They cross-tabulated the responses by the kind of bread, the kind of meat, and the social class of the family thus circularized (e.g., mechanics, businessmen, professional men, farmers, and the university boarding club). Then they studied the students' general eating behavior, carefully recording the composition of various foodstuffs consumed by weight, composition, cost, and nutritional value. Atwater concluded that the people who ate the best were the most affluent and knowledgeable about food. The higher up one went in the American social class system, the better the nutrition.[24] Charles E. Wait of the University of Tennessee conducted the first dietary explicitly focused on southern whites. Three of the four investigations were of students in clubs at the university, and the fourth was of a fairly typical mechanic's family in eastern Tennessee, concluding that these Americans, like those elsewhere, ate "relatively too little of the flesh formers and too much of the fuel ingredients," which was Atwater's overall thesis. Indeed, if anything, this dietary "one-sidedness"—this unhealthiness—was more prevalent in the South than in the North. Hence there did appear to be a common white American diet after all.[25]

Atwater directly supervised two large, ambitious studies of immigrants in New York and Chicago. In the New York study, done with the full cooperation of the New York Association for the Improvement of the Conditions of the Poor, Atwater and his associates studied poor and working-class families on Manhattan's notorious Lower East Side. They studied some twenty-one families, including those of a mechanic, a carpenter, a jeweler, a sailor, a watchman, a sewing woman, a porter, a truckman, and a salesman, among others. In each case, so the assumption went, the occupation mattered for the amount of energy it required and for its relative position in the social order. A category that mattered a good deal to them was ethnicity, or nationality; they referred to it as if it were coequal to, or even more important than, social class or even color of skin in its ability to define the characteristics of a group of people. Atwater and Woods constantly complained that the Irish were much less careful about themselves than such sturdy northern and western European stock as Germans and English people. In an Irish family, for example, the father was unemployed, and the eldest daughter, aged eleven, did all the marketing and housework; the wife was sick and apparently was constantly sneaking a drink. The family spent almost a dollar a day for food, while the father earned, when he was working, no more than fifty dollars a month. "The great trouble here as with so many families in this congested district," Atwater and Woods intoned, "lies in unwise expenditure fully as much as in a limited income," and they groused that people spent money on such extravagant items as bananas and oranges, which had little food value in any event. Not all Irish families were equally disorganized. Thus an Irish truckman was hardworking, conscientious, and decent, but he was always behind on his bills, partly because he was self-employed, but also because his "wife was a very stout woman and drank beer from morning to night, although . . . she would deny it with the can under her apron." Yet they concluded that, despite the father, to get a family as "shiftless and lazy as this one" to reform their ways and spend their food money wisely and economically was almost pointless. In effect, the Irish were an inferior race.[26]

The Germans achieved a different—and better—standard. Thus in one family, the husband and wife were both from Germany. There were four sons and one daughter. "The family was much neater and more thrifty than is usually the case on the crowded East Side," the scientists claimed. The husband earned ten dollars a week as a jeweler, and the two older sons paid the same amount a week, each, to their parents for room and board. Although the family income was not high, they were frugal, hardworking people. They so loved music that they had purchased a piano. Yet they were buying food that was too expensive; if they would only eat less expensive sources of protein, they would be better off. Overall Atwater and Woods were hopeful about the future. They believed that a good campaign of education and persuasion would convince even the most obstinate of the denizens of the lower East Side to change their dietary habits, to purchase more economical sources of food, and thus live a better life. In particular, root vegetables, rather than fresh fruit, and the less expensive cuts of meat and poultry would be the salvation of the immigrant working class once they had adapted to the American standard of nutrition, although they did admit that "among such people there is not only ignorance of the fundamental principles of economy, but a prejudice against economizing."[27]

Atwater and A. P. Bryant, who had studied with him at Wesleyan, conducted the Chicago study through the good offices of the famous settlement house, Hull House. Social reformer Jane Addams cooperated fully. Atwater and Bryant studied fifty dietaries of foreign-born families in Chicago, including five studies of French-Canadians, four of Italians, ten of orthodox Russian Jews, six among liberal or unorthodox Russian Jews, and twenty-five studies among Bohemians, all of whom lived on Chicago's West Side, and they compared the results with three dietaries of American families. Some families resisted the investigators. Many family members were too proud to say how much (or how little) money they had for food or how much it cost, and others would not cooperate in the weighing of food ingested or of wastes. Atwater and Bryant concluded that the poor wasted little. Indeed, generally speaking immigrants were doing pretty well. They ate somewhat better than in the old country and were not extravagant. But they still needed to learn to increase the amount of protein they ate and how to cook "better"—the American, not the old country way.[28]

Clearly the European immigrants were educable. Atwater also ordered up dietaries for persons of color, meaning Mexicans living in New Mexico and African Americans in Alabama and Virginia. Arthur Goss, a professor of chemistry at the New Mexico College of Agriculture and Mechanic Arts, ran two distinct but interrelated studies through his institution's agricultural experiment station. In the first round of dietaries, Goss studied three "typical" Mexican families in the immediate vicinity, two poor families and one of moderate means. In all three families, less food was consumed because less was available than in white families elsewhere in the country, especially with regard to fat, although not with respect to carbohydrates. The Mexicans' diet had little or no meat. They used beans for protein and chilies for flavor and for bowel regularity. Even the Mexican family of moderate means had a very parsimonious diet, having access to little meat—essentially inexpensive fish. Later Goss compared the amount of fat in New Mexico cattle with that of steers raised in Texas, Tennessee, and Maine. Only 0.7% of the New Mexico beef was fat, whereas the comparable percentages for Texas, Tennessee, and Maine were 8.8%, 10.5%, and 18.6%, respectively. He included one more dietary study of a Mexican family and found the same bleak results as with the other impoverished families; theirs was entirely a vegetarian diet, which Atwater and most experts at the time condemned as inadequate.[29]

Atwater arranged for a study of Alabama blacks, conducted through the Tuskegee Normal and Industrial Institute and its president, Booker T. Washington, whose personnel helped find subjects, and with the additional cooperation of the Agricultural and Mechanical College of Alabama at Auburn, whose staff assisted in the data gathering. The research was done in the spring of 1895 and the winter of 1895–96. The investigation included twenty dietaries of eighteen families, some living near Tuskegee, but most living on plantations several miles away. Those families living in proximity to Tuskegee "showed by their improved conditions of living the noteworthy influence of the Institute and of association with people of intelligence and thrift," and the same was true of some plantation families. The rest, however, were very much like the ordinary plantation African Americans "on a large plantation where the mortgage system prevails, and the plane of living is a very low one." Thus the people studied represented the gradation from the lower to the higher grade of living that actually existed "among Alabama negroes, [and] the observations thus help illustrate not only the evils under which the colored people live, but some of the phases through which they are passing in their upward progress." Researchers visited each house or cabin daily for a two-week period, weighed the food consumed, took specimens for analysis, and recorded facts about the people, their dwellings, farms, work habits, and the like. As a class, intoned Atwater and his colleague Charles D. Woods, southern blacks were improvident and lacked ambition because they were ignorant of any better conditions, but they seemed a happy and contented people in any event. Their subjects had enough in their diet as fuel, but, even with the availability of salt pork, which gave them a substantial advantage in protein over the Mexicans, they were still undernourished in this regard, with the amount of protein that Europe's poorest denizens had.[30]

The human nutrition campaign ended in the mid-1900s for several reasons, including Atwater's removal from the scene and enactment in 1906 of the Adams Act, which shut the farming-as-a-business advocates out of the agricultural experiment stations. The studies also came under scientific attack from a new quarter.[31]

Although Atwater was able to establish a so-called dietary standard for white, Anglo-Saxon, Protestant middle-class Americans, his studies of other groups—races and classes—suggested that immigrants and nonwhites ate expensive food as much as possible and ignored Atwater's nostrums, for food was one of their few pleasures in life, either as a habit from the old country or as emotional sustenance in the new. While there were some differences in the averages Atwater's colleagues derived among various groups for percentages of nutrients for fuel and for tissue formation, these could easily be correlated to economic, not racial or cultural, factors. Hence Atwater's approach and his solutions seemed beside the point.[32]

Furthermore, a new school of nutrition arose in the late 1900s and the 1910s to challenge the edifice of technical knowledge that von Voit, Rubner, Atwater, and others had created. A major figure was Elmer V. McCollum, a Kansas farm boy who studied biological chemistry at Yale and was a professor at the Wisconsin Agricultural Experiment Station and the School of Public Health at Johns Hopkins University during his long career. Atwater and his allies stressed the analysis of chemical substances (such as protein) and their physiological effects in the body and the "practical economy" of food, using humans and large mammals as their experimental subjects. McCollum and his colleagues stressed the absence of certain substances in the body and the consequences for health and growth; eventually their work led to the discovery of vitamins in the 1920s and beyond. McCollum studied growth and health in rats, which had a much shorter life cycle and thus provided the possibility for many more experiments than with the Atwater techniques.

Before the 1920s, however, the differences between the Atwater and McCollum approaches were largely those of different disciplinary ideologies, not of fundamental assumptions about the ways in which the world (or, in this case, nature) worked. In *The Newer Knowledge of Nutrition: The Use of Foods for the Preservation of Vitality and Health* (1918), McCollum railed against his rivals. His largest point before the 1920s—indeed, before the later 1930s, practically speaking—was that Atwaterian chemical analysis of foodstuffs was inferior to the biological method which McCollum championed. The chemical method committed one to count calories in the diet. This was clearly inadequate. It did not reveal all the substances crucial to nutrition and health, nor did it make it possible to have a qualitative, as distinct from a merely quantitative, evaluation of foodstuffs and, therefore, of diets. There were large differences in the biological quality of various nutritional substances; thus not all sources of proteins furnished nutrients of equal value. He argued that the old chemical approach, with its emphasis on protein, energy, and digestibility as the criteria of the value of a food mixture, must be abandoned; the chemist could not investigate the quality of what was inside the foodstuff. What mattered was not the relationship of nutrition to thermodynamics, McCollum implied, but rather the role it could play in growth and development.

And that role was only partly chemical. It was, in the main, biological, for it was the *function* of the substances in the food mixtures that mattered, not their chemical composition or structure. There were, in short, growth substances, the lack of which would retard or end development. There were minerals, salts, and what one of the chemical nutritionists, the European Casimier Funk, had first dubbed the "vitamines." McCollum criticized Funk and other chemists for, he charged, they did not understand how these various substances could work, only what their chemical structure was.[33] In the middle and later 1930s and beyond, McCollum and his colleagues could finally identify those vitamins responsible for proper growth and development. Even then, though, it was decades before drug companies could mass-produce these vitamins.[34]

But the middle and later 1930s, let alone the 1940s and the mass production of vitamin pills, represented a far different world, a very different age than the one in which the science of nutrition embraced such seemingly peculiar movements as the NEK, the Atwaterian school of nutrition, and McCollum's protest and call for a truly biological, as distinct from a merely chemical, approach, which started in 1906. From the 1870s to the 1920s Americans from many walks of life acted and spoke as if the taxonomy of natural and social reality operated in a particular way. The structures of reality (and thus their processes as well) were hierarchical, efficient, centralized, three-dimensional, and totally natural. All the elements of the whole were unique; it was as if oil and water could not mix. If McCollum did not, as a rule, make appeals to the idea that each race had its own dietary norm, after the fashion of Richards, Atwater, and their European colleagues and American allies, nevertheless he and his supporters before the 1920s saw the world from a parallel or similar perspective, and no huffing and puffing about the superiority of the biological over the chemical ways of analysis could prove the contrary; indeed, the McCullomites' invocation of biology *ueber alles* simply suggested another theme of the history of science and culture in the period between the 1870s and the 1920s, the exclusionary claims of the various disciplines for territory and turf for their own purposes.

Before the 1920s McCollum and his allies saw humans and animals as machines that, if they grew "right," ate the correct substances that they were supposed to eat according to nature's dictates. This was what made them efficient, what made them grow, what enabled them to become the naturally completed beings that "nature" had programmed them to be. It would be difficult to imagine a more fully articulated parallel construction of thought to the Richards and Atwater versions of human racial nutrition, then, than that of McCollum and his colleagues. That McCollum and his allies were able to do no more than to criticize their professional rivals, and not offer an alternative view of nutrition based on a different taxonomy of reality, simply underscores the point that technical knowledge, like any other product of human culture, is in any ultimate sense necessarily the product of a particular time or era or age and is, therefore, locked in time (although not necessarily restricted to a single national culture, as the close parallels between the general and technical ideas of the German and American nutritionists amply shows). In the 1920s and beyond, as the breeding of hybrid plants and animals proceeded apace in the agricultural and life sciences, so did the newer knowledge of nutrition, but such scientific "progress" depended on a new sense of the order of things, of the meaning of natural and social reality.[35]

PART THREE

Toward an Infinity of Dimensions

Introduction to Chapter 8

Like Zane Miller, Hamilton Cravens has explored the relationships between cultural notions and social science. As Miller examined the ideas of an important turn of the century sociologist on race, here Cravens has studied a group of child psychologists in the interwar years who took positions contrary to orthodox professional opinion on the stability of the intelligence quotient (IQ). By the early 1930s, Cravens argues, almost all child psychologists (and other psychologists as well) agreed that the IQ was fixed at birth. The mavericks Cravens writes about were faculty at the Iowa Child Welfare Research Station (ICWRS), perhaps the leading research center in its field in the world. Cravens insists that notions about stability or nonstability of IQ were largely grounded in the culture's more general ideas about the meaning of group identity for the individual. To what extent was one's identity with a particular group—a class, a race, a gender, or the like—fixed for all time? Or were culture and nature like a suit of clothes, to be put on and taken off virtually at will, as anthropologist Margaret Mead argued in *And Keep Your Powder Dry* (1942), so that the individual could transcend the limits of variation of abilities of the group to which he or she was identified and work within the presumably more generous limits of a more advantaged group? Was it possible, in other words, to think of any and all people as free-standing, autonomous individuals and of the group identities given them by social custom and designation as arbitrary or even wrongheaded?

The Iowa researchers literally stumbled onto their discoveries. Since the early 1920s the Iowa station had had a preschool, which gave faculty and graduate students access to a research population. By the early 1930s researchers had accumulated the longest series of mental and physical measurements on youngsters aged two to six in the country, if not the world, including standard IQ tests at six-month intervals, in the fall and the spring of each school year, so as to assess the impact of the preschool on the children. The director, George D. Stoddard, and Beth Wellman, the professor in charge of managing this data, began to explore why the children seemed to gain, on average, several IQ points from fall to spring, but not from spring to fall; could an enriched school curriculum boost the supposedly inborn, fixed IQ? Wellman and her colleagues published several nursery school studies in which they suggested as much. Interestingly enough, as was the universal practice of their discipline then, they used *group* measurements, even though sometimes it was apparent from their language that they were referring to large IQ changes of *individual* children; the group variations were about five points, more or less what the statisticians and the testers said was the expected variation from test to test, thanks to such factors as the administration of the tests, familiarity with test-taking, selection, and the like. Through a lucky circumstance Stoddard arranged for his colleagues to test their ideas on an enriched curriculum for preschoolers in a state orphanage. This, he believed, would provide an ideal laboratory for the testing of his notions about how the environment (in this case the educational environment) could influence or even permanently alter an individual's basic IQ.

The ICWRS researchers concluded from their studies of children in special preschools and in orphanages that the mental development of very young children could be permanently altered by longtime exposure to environmental influences, positive or negative, by forty or more IQ points, a spectacular finding that their professional colleagues "knew" was as nonsensical as proclaiming that the earth was flat or that the species of the earth were specially created by God. In their (for the time) peculiar notions about group identity for the individual, Cravens argues, the Iowa scientists (who essentially agreed with Mead) departed in certain crucial respects from the culture's commonly understood sense of the meaning of group identity for the individual: namely, from the idea that an individual's possibilities were strictly limited by the possibilities for the group to which that individual belonged. Yet the situation was more complicated than that, showing, as Cravens argues, that the Iowa scientists were prisoners of the era in which they lived (and of its large and controlling assumptions) no less than were their contemporary critics.

The Case of the Manufactured Morons: Science and Social Policy in Two Eras, 1934–1966

8

In 1939, Marie Skodak made an astonishing announcement. This neophyte clinical psychologist at the University of Iowa's Child Welfare Research Station (ICWRS) insisted that under certain circumstances, feeble-minded children could be rescued from their tragic affliction. Skodak pointed to her foster children study where she compared the experiences of two groups of "feeble-minded" youngsters who had been adopted by foster parents whose social, cultural, and educational standing was several rungs on the social ladder higher than that of the children's natural parents. In her research, she traced the mental development of 154 children placed in foster homes at ages of under six months and another 54 so placed under the age of two and a half years. All children came from a state orphanage in Davenport, Iowa. There all tested as feeble-minded on standard intelligence tests. But after a year in their foster homes, the children approximated the mental levels of their adoptive parents, who had well above average intelligence test scores, in the 110–130 range, not that of the feeble-minded or, more specifically, the "morons" or high-grade mental defectives, in the 50–80 range. Thus could children classified as belonging to the group, the feeble-minded, be saved, Skodak insisted.[1]

Skodak's findings were unorthodox indeed in the science of psychology. To her professional colleagues, she might as well have announced that the earth was flat, that the solar system was geocentric, or that God had specially created each species. If there was one maxim on which academic psychologists in the 1930s agreed, it was that intelligence, as expressed in the intelligence quotient (IQ), was fixed at birth and was not capable of being modified during the individual's lifetime.[2]

Skodak had some professional support, especially with teachers and colleagues at the ICWRS; indeed her study derived from their work. And some
of their findings were even more sensational than hers. Especially controversial, even shocking, to many scientists and social workers was her colleague Harold M. Skeels's claim that state orphanages turned their charges
into feeble-minded persons within a year or so of institutional placement.[3]
Nor was this all. Beth L. Wellman, another ICWRS professor, insisted that
the ICWRS preschools promoted permanent gains in IQ for the children
who attended them for any length of time. She even argued that these
changes became permanent among those pupils who stayed in the university's elementary and secondary schools. The environment, Wellman insisted,
was capable of making the IQ gain considerably over a period of months
and years.[4] The Iowa scientists went even further. As their intellectual leader,
the station director George D. Stoddard, repeatedly insisted, if heredity set
the limits of mental growth, nevertheless these were very wide parameters
indeed. Environment could do much to change the child. Early intervention
was the key. The child's mind grew in the early years in important ways,
before he or she ever attended kindergarten or the first grade. The individual
child's mental level became set, Stoddard and his colleagues argued, but long
after conception and birth.[5]

Thus Stoddard and his coworkers contradicted established scientific opinion. And they sharply attacked conventional therapeutic practice for "feeble-
minded" children. They proposed a spectacular and dramatic overturn of
established professional opinion at every point. The evidence that they ad-
duced of large-scale gains and losses in the IQ of very young children, in
response to long-term environment pressures, profoundly shocked and
aroused some of the nation's leading experts in psychometrics, or the field
of mental measurement. It also suggested that perhaps something was rotten
in Denmark—that, in short, institutions established for children, namely the
public schools and state orphanages, limited, perhaps devastatingly so, the
potentialities that children had for mental growth and therefore for a wholesome social existence in American culture.

But there was more. The technical findings that Skodak and her teachers
proffered to their professional colleagues and lay audiences also contradicted
fundamental assumptions that most Americans shared in the interwar years
concerning the structures and processes of the U.S. social order. In the mainstream view, a rising tide carried all boats but did not rearrange their hierarchical relations; each group in the national population had its own place
in the larger society, whether that was at office, store, or factory, doing the
laundry or cooking in the kitchen, living in the barrio and following the
harvest, or drinking from "colored only" water fountains and riding at the

back of the bus. This group determinism implicit and explicit in the culture at large, which was itself reflected in well-known gender, class, and racial or ethnic relations of the interwar years, also permeated the intellectual and perceptual structures and processes of science, in this case the technical knowledge of the psychological scientist. That the Iowa scientists took on such a powerful battery of assumptions, and the fate they experienced for their trouble, forms an interesting and revealing case study in the history of technical knowledge in American culture.

The Iowa Child Welfare Research Station was established in April 1917. It was founded thanks to the long-term lobbying efforts of a coalition of progressive-minded women activists to create a research institute whose staff would investigate all aspects of child life through the scientific method. The activists assumed that science would save children. The ICWRS was the first such broadly gauged scientific research institute devoted entirely to children in North America, and probably in the entire world. It was also the only such institute in North America, if not the world, that owed its existence to activist sociopolitical reform, a point that its staff never forgot until after World War II.

In the 1920s the Laura Spelman Rockefeller Memorial disbursed several million dollars to create a professional scientific subculture for the yet-to-emerge science of the child, or child development, including establishing several research centers, as at Minnesota, Columbia, Berkeley, and Toronto, extending the ICWRS and expanding Arnold Gesell's child study center at Yale, as well as a larger number of institutes at which the new science of child development would be disseminated to presumably eager parents in so-called parent education programs. The LSRM also played financial midwife to the founding of three journals, a scientific society, and a postdoctoral research fellow program, and even subvened the publication of a popular magazine, *Parents Magazine*. In the crystallizing profession of child development, the ICWRS was the undisputed leader, the most important single institute. It was the only one that attempted to cover all phases of the subject. It trained most of the first generation of doctoral students, and its staff was more productive than that of any other institute in the 1920s. Being the first institute in the field mattered; it enabled the ICWRS to maintain its reputation as its members became the chief dissenting minority in the profession and discipline they had founded.[6]

As the professional subculture of child development had taken shape in the 1920s, so the discipline or science of child development began to crystallize in the early 1930s. Most child developmentalists now accepted two major theories as fundamental to the science: the notion of the fixed IQ and the maturation theory. As John E. Anderson and Florence L. Goodenough of the Minnesota Institute of Child Welfare put it in their well-known text, development in all sentient beings was lawful and followed certain regularities. According to the notion of the fixed IQ, the individual was born with a certain IQ, fully within the range of variation of the group to which he or she belonged in nature and society, and nothing could change it. If the IQ scores of a person varied, the causes were various errors or irregularities in the administration of intelligence tests or, perhaps, in the lack of competence of those interpreting test results. Advocates of the maturation theory insisted that growth itself was predetermined by the individual's inheritance. As the individual's particular innate neural organizations matured, environmental stimuli would awaken them, that is, arouse their fairly well-fixed patterns of behavior. Heredity and environment worked together in a larger symbiotic relationship, as could have been assumed from the concurrent resolution of the heredity-environment controversy in the natural and social sciences, but the resolution of that controversy meant a continuation of the subordination of culture to nature, at least insofar as the purview or turf of the natural sciences was concerned. More important than the culture-nature dichotomy, however, was the issue of determinism versus indeterminism; most developmentalists, like most scientists, ultimately were determinists.[7] And what a comforting brace of scientific theorems to those firmly committed to the social class and caste order of contemporary America. Technical knowledge seemed to reflect and to sustain deeper assumptions about social mobility and stratification that were fully integrated into social structures and cultural perceptions of the nation at large. Because the cultural assumptions and social practices existed before the technical knowledge in the case at hand, it may be fairly assumed as to which was source and which was result. The group determinism of the culture at large thus helped inspire and sustain technical knowledge about human development.

Thus did developmental science contain within it the larger culture's ideas about the meaning of group identity for the individual and, indeed, about the autonomy, or lack thereof, of the individual in an allegedly free and democratic society. Alice M. Leahy's monograph, *Nature-Nurture and Intelligence* (1935) ably represented the majority point of view both within the larger culture and within her scientific profession. Leahy had studied with Goodenough and Anderson at the Minnesota Institute of Child Welfare, which was an important bastion of hereditarian thinking in the field. In turn Goodenough and Anderson had worked with those important architects of mental testing and of the hereditarian and deterministic interpretation of such testing, Goodenough with Lewis M. Terman, who devised the Stanford-Binet, and Anderson with Robert M. Yerkes, who had invented and administered the United States Army's mental testing program during World War I. Goodenough and Anderson developed more than ideas on the fixity of IQ. They constructed measurements and, therefore, notions of the environment too. They created an occupational scale, drawn from the 1920 federal census. They devised a hierarchy of eight groups, the most prestigious white collar occupations at the top, with those at the bottom requiring the least skill and education. To assess the home environment they used a sociologist's rating scale that assigned different weights to various types of material possessions in the home. Thus the higher the occupational and educational level of the parents, and the more highly ranked material possessions they owned, the higher the social status and, therefore, argued Goodenough and Anderson, the innate intelligence of the parents and their relatives—certainly their children, but all to whom each parent was biologically related.[8]

Leahy brought these assumptions to her study. She compared the resemblance of parents and children, the one group of children natural, the other adopted. Her central thesis was the power of biological mechanisms; but she did not dismiss the power of environment. Rather she insisted that what biology did not determine, environment did, thus spinning out a theory of mental growth and development that was profoundly *deterministic*. But there was more. And it mattered even more. To buttress her argument that children always resembled their natural parents more than their adoptive ones, she deployed the measurement that was standard in the scientific industry in which she worked: group measurements. Not once did she measure one individual from one group against another from the other, nor did she track single individuals over time. In her view, all portraits were those of a group. It did not occur to her or, for that matter, to almost any other scientist of that era, to regard individual measurements as valid. Group measurements were what mattered, for the good and simple reason that they, not individual measurements, reflected natural and social reality. Here again, notions from the larger culture informed technical knowledge.

How, let alone why, Leahy and virtually every other scientist of that era assumed that group measurements were real and individual ones were not is too complex a problem to discuss here, but obviously the single most important culprit was the ideas and practices of the culture itself, as distinct from the science of psychology. When mental testers had used group averages as their standard for determining whether the IQ was fixed or not, said measurements always appeared to demonstrate very little, if any, variation in IQ in group averages from test to test, even of the same groups of individuals. It is important to grasp the point that almost no psychometrician tracked the scores of individual children over time; such would have been inconceivable because both the larger culture and the disciplinary discourse of which they partook "taught" them to believe that individuals could exist only within a group—and with a group identity.[9]

Furthermore, the Binet test had been designed so that half of the children tested for any age group would pass, one-fourth would fail, and one-fourth would excel. Here again was the interest in group, not individual, measurements. Furthermore, Leahy studied children ranging in age from five to fifteen years—a period in which there was, comparatively speaking, much stability in the IQ simply because of the dynamics of human mental growth. These were, to be sure, assumptions that all in the field took for granted. And they were assumptions that teachers had been using for many years in setting the curve in their classes. And beyond that, they were notions that were implicit and explicit in social customs and cultural notions throughout American society and culture.

Although Bird T. Baldwin, the first ICWRS director, vigorously attacked the hereditarian and fatalistic interpretation of mental test results in the last year or two of his life, it was really Stoddard, his successor, who developed a full-blown program critical of the mainstream deterministic theories of child development. Stoddard combined political activism and scientific research that both centered on the still-inchoate notion of the autonomous individual rather than the group determinism so pervasive among child developmentalists. A left-liberal supporter of the New Deal, Stoddard believed in the power of education to cure most, if not all, of society's ills. During his early years with the ICWRS, he spent much time arguing before a variety of public forums for the creation of a national system of public nursery schools or preschools for children from the ages of two to five or six. This would help stamp out poverty and poor health and would benefit the masses. It would help stimulate young children to be better students and more constructive and cooperative citizens.[10]

Stoddard developed the scientific side of his program in a piecemeal fashion. He inherited a staff whose work tended to stress, without much discussion or perhaps even awareness of the point, that the individual, not the group, was the point of departure for all investigations. As early as 1928, for instance, Beth L. Wellman, a psychometrician at the ICWRS, was synthesizing the records of the children in ICWRS preschools to ascertain the influence of the preschools on mental growth and development of young children. She had no self-conscious notions then of the individual as distinct from the group. Rather she tracked the performance of individuals and used group measurements to present her data. By the middle 1930s, Wellman and her coworkers had examined the test scores of the 1,333 children who had attended the ICWRS preschools and the university elementary school and high school. She insisted that all children gained in IQ points while attending these schools, for the mean IQ on the first IQ test for all 1,333 children was 100, or average, according to the way the Binet test was constructed; for 1,027 children who stayed in the system long enough to take a second and third IQ test, the mean IQ was 119, and for 574 children who took a fourth, fifth, sixth, and seventh IQ test, the mean IQ was 124 points. She was using, of course, a group measurement; such made sense in her field, not to mention in the larger culture.

But Wellman also referred to the performance of individual children, as when, for instance, she discussed which groups of children gained the most and the least while attending the university schools. While she classified these groups according to the average amount of gain they made, with those having the lowest initial scores making the largest gains and those with the highest starting IQs the smallest gains, note that these were groups that she had assembled from the data itself and only for the purpose of representing the data. She had to examine the records of individuals to construct the groups, in other words. Her general conclusion, that the enriched environment of the preschools was continued in the elementary and high schools and explained these startling gains in IQ scores, was an analysis of the performance of *individuals.* But she expressed said performances as if they were measurements of groups, not individuals, even when it was clear, at least in retrospect, from reading her text that the reality that had caught her eye was the behavior of a relatively large number of single individuals within a large sample.[11]

Two years later Wellman published another report in a similar vein. She presented information on three different groups of children. One had transferred from the university schools after attending the preschools. Another had continued all the way through the university schools and had graduated from high school. And a third group did not attend any schools at the university. She found that those who transferred out had gained, on the average, about nine IQ points while in the university schools and that those who had never attended them experienced an average loss of about one IQ point, whereas those continuously enrolled gained an average of seventeen IQ points! Thus Wellman insisted that the environment—that of the preschool—was crucial to mental development in young children. Obviously this was grist for Stoddard's political mill.[12]

It is important to note that Wellman recognized and recorded the performances of individuals in mental growth but presented such data as if they were group phenomena, that is, only in group measurements. Her national professional colleagues naturally assumed she was discussing group phenomena—and she probably thought so also. Of course she was doing no such thing. And therein lay the rub. To this point her national colleagues had found but meager gains in IQ scores—five points or less on children retested if the initial test was at age six or older—because they paid attention only to the performance of groups of individuals, not of the individuals themselves.

Wellman and her colleagues made bold, if not incautious, claims about their work. Thus in one project completed with a student, Wellman challenged the mainstream argument that cultural status as measured by such factors as occupation of parents and quantity of material possessions had any meaningful relationship to IQ standing. Again she discussed individual performances as if they were group phenomena, replete with group measurements. Wellman could certainly wring a political message from her material. Her thesis was that the preschool experience was crucial in boosting the IQ of children, and the gains there were or tended to be permanent. And she came close to insisting that culture and nature were forces whose power could be overridden in individual cases with the right kind of early intervention and education. The "amount of gain made [by the preschoolers so studied in this investigation] was contingent upon the intelligence level attained when the child began his preschool career, irrespective of his cultural status," she concluded. This was a profoundly subversive message, for it assaulted widespread contemporary assumptions about the American social system contained within the larger culture itself. Simply put, it suggested that group identity for the individual could be overridden via educational and other environmental forces, so that there was no longer any excuse for poverty, racism, or for discrimination against any stigmatized group in society whatsoever.[13]

Yet neither then nor later could Wellman articulate the important point that she was discussing the experiences of individuals, not groups, even when it was clear from the texts she published that this was so. Again and again she referred to group measurements, yet her anecdotal discussions were all about individuals. That was how powerful was the notion of that age that the ultimate reality was the group, not the individual; the individual was only an example of a tendency within the group to which the individual belonged or was thought to belong. Obviously these were peculiar, even oddish, developments.[14]

What became known in the later thirties and early forties as the Iowa orphanage studies began almost by chance. In the early 1930s a prominent couple in the state adopted a child from the Iowa Soldier's Orphans Home. Subsequently the child was diagnosed as severely mentally retarded. The couple threatened to sue the State of Iowa through its Board of Control, which governed the Davenport home. To help avert the suit, the board's members agreed to monitor orphans more carefully henceforth. They turned to Stoddard for advice. He seized the chance to enlarge ICWRS access to research populations. He persuaded the board's members to accept a program in which a clinical psychologist appointed at the ICWRS and the home would be responsible for giving IQ tests to all inmates and for recommending appropriate placement. In return, ICWRS staff could conduct research there. Appointed was Harold M. Skeels, who had studied with Wellman and Stoddard; his specialty in psychology was intelligence in very young children. The program officially began in February 1934. Henceforth all inmates would have a Binet test in addition to the usual physical, medical, and dental examinations.

Over the next year Skeels gave Binet tests to seventy-three children prior to their permanent adoption by foster parents. These children had been placed with their foster parents at six months or less, and their mean age at final adoption was about thirty months. Like most professionals in clinical psychology, Skeels assumed that the foster children would more nearly resemble their natural than their foster parents in IQ scores. With this group, the mean IQ of the natural mothers was 84 points, definitely marginal between dull-normal and "feeble-minded," as the mentally retarded were then known; only 10 percent scored more than 100 points, or normal, on the Binet, and 38 percent scored lower than 80 points, definitely feeble-minded. The IQs of the natural fathers were as unavailable in these instances as they almost always were. But it seemed clear that, as was usual with orphans in state institutions, both natural parents were at the bottom of the occupational and educational hierarchies that Goodenough and Anderson, among others in the mainstream, had developed. What was known about the fathers of the seventy-three children was that they were indeed as unpromising genetic material as the mothers.

Yet Skeels's results were so contrary to normal professional expectations as to make him and his colleagues rub their eyes in amazement. The seventy-three foster children had a mean IQ score of 115.3 points, definitely above a normal score of 100, well above that of their natural mothers, and clearly in the range that clinicians and child psychologists, including Goodenough and Anderson, thought reserved for the offspring of middle-class, well-educated parents of northwestern European ancestry. According to the calculus that Goodenough and Anderson advanced, those foster children should have had a mean IQ somewhere in the low 80s—82.5 was the mean that Goodenough suggested for one study—because of the low status of their natural parents, a reflection of their unpromising genetic material.

Skeels insisted that his results put Wellman's data on gains in IQ of middle-class preschoolers in a new light. Children from unpromising cultural and biological backgrounds could nevertheless make dramatic gains in IQ scores if proper intervention took place early enough in the life cycle. As he said in a public address, tragic errors were being made in placement of children in state institutions almost daily. He had already found forty children with normal IQs at the state home for the feeble-minded, and two hundred feeble-minded children in other state facilities where they plainly did not belong.[15]

At the same time Stoddard established an ambitious research project for the orphanage. It lasted for three years, from October 1934 to June 1937. Stoddard was its major architect, although Wellman and Skeels also contributed to the planning. The project's design appeared simple, at least at first blush. Its overall purpose was to explore the influence of the nice middle-class curriculum of the ICWRS preschools to see if it could improve the prospects, social and academic, of the poor oppressed children in a typical state orphanage. A control and an experimental group were selected from the orphanage's population. Each had twenty-one members. Each individual child was matched as closely as possible with a child in the other group. Those in the control group were carefully observed and their developmental history carefully monitored. Those in the experimental group were put into a special all-day preschool, modeled on the ICWRS preschool in every possible way in terms of equipment, curriculum, and the like. Again, each child's development was carefully observed and recorded. Built into the experiment was the assumption, borrowed from Wellman's work, that the preschool would make a large difference in the lives of the children who attended. It would intervene early enough in the life cycle to transform the children in the experimental group from their wretched past to a promising future. A methodological problem for which there was no solution was that over the course of the project, as individual children were adopted, they left the project and were replaced with others as carefully matched with them (and their opposite member in the other group) as possible. The mean IQs for the groups were close, 82.3 for the experimental group, 81.2 for the control.

Matters went awry from the start. The nursery school teachers found that their charges were almost entirely antisocial in their behavior. It took most of the first year merely to teach the children to be toilet trained, to clean and dress themselves, to treat the equipment (and one another) with respect—in short, to simulate the behavior of well-scrubbed middle-class children with educated parents, not that of the abandoned and desperately antisocial lower-class waifs that in reality they were. In the preschool's second year, problems of social adjustment diminished, and the teachers slowly introduced the traditional curriculum and routines from the ICWRS preschools. As the children eventually learned how to live as normal middle-class children, the teachers could then concentrate on helping each individual child become emotionally, socially, and cognitively developed. Those socialized who remained helped break in the replacements, so that by the second spring the teachers judged that theirs was, at long last, a reasonably "normal" preschool. The teachers understood that they had effected a miracle that they had neither expected nor been prepared to produce. The project was terminated in 1937 for budgetary reasons; the fortunes of "feeble-minded" children were not a high priority with state officials anxious to genuflect before taxpayer outrage at such lavish expenditures.

In 1938 Skeels, together with several colleagues, published a monograph on the project. It was the project's findings with regard to mental development that sparked the most comment and controversy. And the facts were tricky indeed. No child remained in either group longer than twenty months, thanks to adoptions; and for technical reasons, the scores of eleven experimental and five control children were discarded. Over the longest period of residency, twenty months, the mean gain, Skeels argued, was 4.6 IQ points for children in the experimental group, whereas the mean loss for those in the control group was, again, 4.6 IQ points. That astonished Skeels and his colleagues, who expected a larger gain in IQ among the experimental children than indeed occurred.

But the most startling discovery that Skeels and his associates made was also unexpected by those involved and was not, strictly speaking, the result of a controlled experiment. That discovery had nothing to do with the children in the experimental group. Skeels and his associates found that the longer many or even most of the children in the control group remained in the orphanage, the more likely it was that their final IQ scores as individuals would move downward, toward the high feeble-minded classification, 70 to 79 IQ points, or even lower. Thus it was not the preschool but the orphanage that was the more influential institution in the children's lives. The orphanage could not be regarded as a benign or constructive or positive institution. It was destructive of its hapless charges in the extreme.

Skeels and his associates were caught between two contradictory levels of discourse. Thus they spoke and wrote as if they agreed with their national colleagues that groups existed in nature and society and that individuals were but manifestations of such group reality. And indeed they did believe that, as their constant references to group measurements made clear, as in their contrast between the mean gains and losses in IQ (4.6 points) for the preschool and control children. Yet they also thought of individual behavior in ways sharply contradictory to the received wisdom of the age. They shifted their level of discourse from group to individual without a language or a method for it. They insisted, for example, that the largest losses in IQ points were for those children with the highest initial IQ—a startling claim indeed, if one believed that the IQ was fixed at birth and that these losses ranged from 16 to 40 IQ points, a truly staggering and unique finding for that day, but also a statement about individual, not group, performance. They insisted that twenty-six of forty-eight pupils for whom such indices could be calculated experienced a mean loss of 16.2 IQ points; obviously, Skeels and his associates were thinking of individuals as they assembled their data and then calculating group measurements without understanding the implications of their statements.[16]

Nor was this all. Adventitiously Skeels found more material for his developing argument when the circumstances of two retarded infants routinely transferred from the Davenport home to the state home for the feeble-minded in Glenwood came to his attention. At ages thirteen and sixteen months, respectively, the two infants had tested at 46 and 35 points on the Binet scale. Because the Glenwood institution was overcrowded, the infants were placed on a ward with female inmates ranging in chronological age from eighteen to fifty years and from five to nine in mental age. Six months later Skeels was on a routine visit to Glenwood, when he discovered that the two infants so situated exhibited much improved behavior—they sat up without assistance, they played with toys, and they responded to adults in ways unheard of for children with their alleged dismal IQ scores.

Initially Skeels wondered if he had not erred in his original diagnosis of the two infants. Retesting with a special version of the Binet produced yet another surprise. The infants, now eighteen and twenty-two months old, tested at 77 and 87 points, gains of 31 and 52 points, respectively. Twenty months later, the younger child tested at 95 and the older at 92 IQ points, an additional rise of 18 and 6 points, respectively. Now that the children tested in the normal range, Skeels ordered their transfer back to the Davenport home, where in due course they were permanently adopted. Skeels explained this dramatic change as the result of the special circumstances on the ward at the Glenwood institution, in which the intensive attention and care the children received from the adult inmates helped them achieve a normal level of mental development. Such would not have continued beyond the point at which they surpassed their nannies, Skeels argued, but by then their case for adoption made their situation far more promising than had initially been the case.

To test this theory even further, Skeels arranged with the superintendent of the Glenwood institution to transfer another thirteen feeble-minded children from Davenport to Glenwood to be cared for under the same circumstances as the two infants. These children were too severely retarded for adoption; their mean chronological age was about twenty months, and their mean IQ was about 65 points, with a range of 35 to 89 IQ. And over a period of eighteen months they were compared with a control group of children at the Davenport orphanage who were about the same age and borderline normal. The results of this study were as shocking as those of the larger orphanage project. The children in the experimental group made an average gain of about 28 IQ points, and those in the control group experienced a mean loss of 26 points.

Put another way, the Davenport orphanage seemed to be manufacturing morons, or high-grade mental defectives, out of human beings who just might have had radically different prospects in life were they not in such an environment. If Skeels presented his data in terms of group measurements, nevertheless he and his colleagues were thinking also in terms of individual performance; sometimes they spoke and acted as if individuals, not groups, were the ultimate stuff of natural and social reality, and at other times the exact reverse was the case. Obviously there was a conflict between what the larger culture and his own technical knowledge "told" Skeels; this was, in short, a spectacular instance of cognitive dissonance. Such considerations did not engage Skeels' attention, however. Ultimately Skeels drew, as was his wont as a clinical psychologist, a practical rather than a theoretical point: he suggested intensive early intervention for such damaged children, meaning many more adults than children, and meaning too (probably) that each individual only tentatively belonged to whatever group society and nature assigned that person.[17]

In the inevitable controversy that arose over the Iowa work within the fields of psychology and child development in the later 1930s and early 1940s, there were the Iowa scientists on the one side and Lewis M. Terman, inventor of the Stanford-Binet test, and his allies, notably Goodenough and Anderson, on the other. Perhaps the most intense debates occurred in 1940 and 1941. These events have been covered in some considerable detail elsewhere, especially in published form in psychologist Henry L. Minton's fine biography of Terman and also in my book on the Iowa Station. There is no need to replicate that detail here.[18]

Terman had served as general editor of a volume or yearbook of studies on the relative importance of heredity and environment in human intelligence for the National Society for the Study of Education (NSSE) in 1928. Such enabled him to showcase the work that his protégés were doing in the area. Hence it was hardly astonishing that Stoddard would propose that he do the same as the work of his colleagues on mental measurement came out. That yearbook came out in early 1940, after four years of intense—and sometimes tense—interaction among the disputants on both sides of the issue. Terman and Goodenough recruited a majority of the volume's contributors, all deeply opposed to the Iowa work, and all supported that idea that the IQ was fixed and that maturation was predetermined.[19]

Yet the evidence the thirty-eight contributors presented in slightly more than twenty original studies by no means destroyed the arguments that Stoddard and his other ICWRS colleagues had made. Apart from Wellman's defense of her work with the preschools, there were nine separate investigations of mental development in nursery schools. The authors of eight declared themselves opposed to the Iowa studies. Yet in every case the authors had replicated Wellman's methods in retesting, and all found some kind of mean gain in IQ points, with some noting considerable changes in the scores of particular individuals. The authors of one of the nine studies agreed with Wellman; the others, including Wellman's most severe critic, Goodenough, found mean gains ranging from two to seven points in a year. Those who disagreed with the Iowa scientists simply dismissed their findings as inconclusive.

Indeed, the Iowa scientists and their critics were essentially talking past one another. The Iowa scientists were caught between using the accepted professional discourse of group behavior and referring in fact to the heretofore unplumbed behavior of individuals, which was totally different. Their critics, such as Terman and Goodenough, used either the technical language of group statistics, such as means, medians, and modes, or child development's professional discourse of group description and analysis, which amounted to the same thing. When Goodenough, for example, discussed the Iowans' statistical errors as including sampling problems, inflated Ns, and the like, she was excoriating them for not sticking to measurements of groups. In a real sense, the Iowa scientists were interested in variation among individuals, whereas their critics were interested solely in group characteristics.[20]

Yet sampling errors and inflated Ns were meaningless criticisms if one were concerned with tracking single individuals over time, which is what the Iowans were doing without quite realizing it. That they did not fully grasp what they were doing was completely evident from their texts and correspondence, in which they constantly referred to changes that individual children made but did not, before 1942, after the controversy was essentially over, make any formal representations of said tracking of individuals. And the professional and disciplinary discourses of psychological science were no help at all with regard to the representation of the kind of individuality upon which the Iowa scientists had stumbled. For some time in the 1940s one of Terman's colleagues, Quinn McNemar, had carried on a technical debate with the Iowa scientists over their data, primarily focusing on their rather unsophisticated understanding of statistics. McNemar's original critique, published in 1940, did a great deal to damage the Iowa findings' credibility. Yet the Iowa scientists turned over their data to McNemar, and in the evolving technical discussion it became perfectly clear that neither the Iowa scientists nor their critics could address the phenomena of individuality in the same way that statistical methods, theories, and assumptions could facilitate descriptions and explanations of groups.[21]

And, indeed, the controversy was almost over before it began. As Goodenough wrote to Terman on the eve of the publication of the NSSE volume, already she thought that the Iowans' work was "going the way of ESP and I strongly suspect that within a couple of years from now the whole matter will have passed into the same limbo of lost memories."[22]

Goodenough's prescience was more accurate than even she could have imagined. By 1942 the Iowans were in full retreat. Stoddard left for another position, Skeels went into the service, and the new director, Robert R. Sears, while sympathetic to the program in mental development, was infinitely more interested in starting a different kind of research program in the study of child personality. Terman and his allies had been able to publish critiques that damned the Iowa work before the profession as statistically flawed, methodologically sloppy, and woollyheaded in its practical or therapeutic recommendations. This is not to say that most psychologists and child development researchers were interested in the debate. Far from it. Most assumed that the IQ was fixed. They wondered, if they bothered to notice, what all the fuss was about. From the perspective of that age, the ICWRS work made little sense. A handful of clinical psychologists interested in the same issues as the ICWRS scientists did support their work. But they were given no more than a respectful hearing, with their ideas ascribed to those impractical do-gooders in clinical psychology. Real experimental psychologists knew better. American culture and this particular piece of technical knowledge were in perfect harmony. And that was that.

And there the issue rested until the early sixties. In 1961 J. McVicker Hunt, a child psychologist who had always fused science and advocacy in his career, published a seminal book, *Intelligence and Experience,* in which he reassessed the classical work in his field and, overall, found it wanting. In particular he subjected the notions of the fixed IQ and of automatic maturation to tough criticism. In so doing he also cast aside the group determinism of that prior age, insisting that the individual, not the group, was and should be the center of all scientific analysis and policy proscription concerning children in American society. He linked much research done in the field since the later 1940s that corroborated, directly or otherwise, the work that the ICWRS workers did in mental development. Thus research on early mental development showed that many individual young children did experience large fluctuations and changes in IQ in response to all manner of postnatal events in their milieu. Work on maturation in animals suggested that maturation was neither automatic nor presaged, which, given the power of the evolutionary analogy and the pure-science ideal among natural scientists, helped legitimate the notion that perhaps development in children more nearly followed the dictates of "learning" in the environment rather than the environment merely awakening genetically predetermined responses. The implications of Hunt's arguments for public and social policy were clear: early intervention was not only scientifically justified; it was a moral imperative.

In 1964 psychologist Benjamin Bloom made Hunt's argument even stronger. Bloom insisted that half of the variation possible in mental powers was determined by the age of four, and therefore it was essential to intervene early in the life cycle to rescue the child who would otherwise be "at risk" from external social circumstances and forces. In the summer of 1965, the federal government began supporting day care centers around the nation for so-called at risk children, under the broad authority of the Economic Opportunity Act that the Congress had enacted the previous year. In that first summer of what became known as Head Start, more than half a million children were taken care of in more than eleven thousand day care centers. While cognitive development was not specifically mandated by the law, it became one of the major objectives of the Head Start program, and it is indeed that part of Head Start that remains generally accepted throughout the political spectrum in our own time. And Head Start was predicated on the assumption that the starting point for the uplift of society was the individual.[23]

In 1966 Skeels published a follow-up to his 1942 Glenwood study. He had left the ICWRS in 1942 for wartime service and, after World War II, pursued a career as a clinical psychologist in the federal government. In the early 1960s he began tracking down all twenty-five subjects in the study that he did of the inmates at the Glenwood institution for the feeble-minded. This study had become one of the special targets of Terman and his allies when it was published in the early forties, with Terman derisively dismissing it as the "moron nursemaid" study because of Skeels's claims that the older female inmates rescued the retarded infants from feeble-mindedness. Skeels found every individual of the thirteen in the experimental group and the twelve in the control group. The thirteen in the experimental group, nurtured as they were by the "moron nursemaids" as very young infants, nevertheless grew up to have normal lives on a par with those of their adoptive, not their natural, parents. The life experiences of the twelve who remained in the Davenport orphanage could not have been more different. They were constantly in trouble with the authorities, their school history was dismal, and their employment history spotty at best. Even in personal and family life the contrast continued, with almost all in the experimental group having normal family lives, being responsible parents, and the like, and almost none of those in the control group having anything approaching an emotionally normal family and personal life. Given the fact that it was the thirteen who had originally been diagnosed as feeble-minded, and the twelve in the control group who had tested as normal (if marginally so), the conclusion was inescapable: early intervention mattered a good deal.[24]

And it was perfectly obvious in Skeels's account, as in the field's new literature of the sixties, that the focus of the field had silently but massively changed, from a belief that one's group identity set one's possibilities to a radically different assumption, that the individual could achieve a full and normal life if allowed to develop in a wholesome and constructive manner. Skeels's work was readily accepted as a part of the new ideas of its age. No longer was he a heretic but merely another laborer in the vineyard. Obviously there had been a seachange in attitudes in the larger culture as well as in his profession, for the model of development of the interwar years had been turned on its head.

To account for how, let alone why, this transformation occurred would require infinitely more space than is available here. This much may be suggested. When the Iowa scientists published their original work in the 1930s and early 1940s, they worked in an age in which the taxonomy of natural and social reality—that age's sense of the order of things—made it virtually impossible to see or perceive, unless forced to, that an individual could behave outside the norms or limits of the group to which nature and society had assigned that person. In particular, as I have argued elsewhere, that age, stretching from the 1920s to the 1950s, assumed a taxonomy of reality in which the individual was merely a manifestation of incredibly complex and intricate *group* processes. Put more abstractly, the whole was then understood to be greater than or different from the parts, which were in turn said to be distinct yet interrelated. This was a kind of holistic determinism that succeeded the better known reductionist determinism of the later nineteenth and early twentieth centuries in American (and western) culture. In the fifties and beyond, however, it has become increasingly obvious that we live in a radically different kind of age, in which the taxonomy of natural and social reality assumes an infinity of dimensions, portions, shapes, and parts, an individuation of relationships that is virtually incalculable and, in political and social policy discourse, an age of individualism and, more specifically, of the victimization and liberation of the individual against the "system" or the larger whole. In such an age it suddenly "made sense" to view children as individuals, not as members of iron cage–like groups. That would also explain why the simple empirical discovery made by the Iowa scientists, and confirmed in fact but not in interpretation by others in the late thirties and early forties, seemed so elusive and difficult to express then although so simple and obvious to us in our own time; quite evidently, the ground has shifted, and dramatically so, since the interwar years, and it was the larger notions of the culture about the structures and processes of natural and social reality themselves, far more than the knowledge that the technicians had invented or reinvented, that told the tale.[25]

Introduction to Chapter 9

Here, in the volume's penultimate essay, Robert B. Fairbanks addresses yet another aspect of the functioning of technical knowledge in American culture: the uses that representatives of vested interests make of technical knowledge in pursuit of their own goals. Fairbanks takes up the fascinating phenomenon of airports and their relationships to urban commercial, corporate, and especially municipal interests in pre– and post–World War II America. Focusing on the Dallas–Fort Worth area, the part of the nation that gave the world the deliciously garish term "metroplex," he finds differences in definitions of "true" residential and influence units as the explanation for the repeated contentiousness between Dallas and Fort Worth and between each city and the federal aviation bureaucracy.

Fairbanks argues against technological determinism and for the thesis that it was different notions and definitions of what constituted society and economy that helped shape the wrangling over airport location in Dallas and Fort Worth. In the interwar years, Americans from most walks of life acted and talked as if the basic unit of society was an organic community, located in time and space. Thus the early lead that Dallas business and political interests built up in having an airport for their city—Love Field—reflected that definition of social reality. That Love Field was also built essentially with local funds, as distinct from federal monies, buttressed that perspective. When Fort Worth's leading citizens began agitating for their own airport, or for one that might threaten to subsume Love Field, rivalry broke out. So long as the two camps acted and spoke as if their two communities were organic, self-contained entities, distinct from one another, cooperation on an airport for the Dallas–Fort Worth area made little sense, for the very good reason that Dallas and Fort Worth did not seem to make a coherent unit to contemporaries.

In the 1950s and early 1960s, however, the federal government entered the dispute as a major player, arguing that Dallas and Fort Worth had enough in common for a regional airport. Although this new perspective can be viewed as the imposition of a national view by federal bureaucrats on fractious, competitive locals, and Fairbanks suggests that this is a helpful explanation, he also insists that more was involved. What resolved this persistent set of disputes, what allowed the rivals to agree on creating the Dallas–Fort Worth International Airport, was the idea of a metroplex. A public relations firm first articulated the notion of a metroplex in 1971. And that jargonic term was symptomatic of a new definition of social reality and society. The basic unit of society was not an organic community but the individual ready to pursue the imperatives of a market economy. With that new point of view, the idea of a metropolitan Dallas–Fort Worth airport made sense in ways it could not have done before. In this case agreement on what constituted the real sociopolitical, economic, and cultural arena resolved long-standing confusions and fractiousness. Fairbanks's essay thus parallels the contribution by Cravens on IQ. Both find the locus for change, for understanding, and for ultimate public action to be the result of redefinition. And the redefinition that they discover is the same: the holistic early twentieth century was succeeded by the individuated later twentieth century. Thus was technical knowledge, whether of the wandering IQ or of the urban metroplex, informed and shaped by larger cultural notions characterizing the taxonomy of reality of the age in which they existed.

ROBERT B. FAIRBANKS

Responding to the Airplane: Urban Rivalry, Metropolitan Regionalism, and Airport Development in Dallas, 1927–1965

9

From today's perspective, Dallas's refusal to join with Fort Worth and cooperate with the federal government during the 1940s and 1950s to develop a regional airport midway between the two north Texas cities seems immensely foolish and shortsighted. The Dallas–Fort Worth International Airport is one of the nation's largest and helps anchor the eleven-county metroplex, covering 8,360 square miles and inhabited by more than three million Texans. The tradition of urban rivalry and bickering between Dallas and its nearby urban rival, Fort Worth, some thirty miles west of Big D, seems to confirm an image of parochial and petty Dallas leaders unable to see the importance of an airport that might help create a super-region. But closer examination of the Dallas airport story suggests that such a simple explanation of Dallas actions may be inappropriate. For the business and political leaders who fought the proposed regional airport while helping to usher in an era of unprecedented prosperity to the metropolis based their decision to fight the regional airport in part on the advice of outside planners, nationally respected experts. And the recommendations of these expert planners were influenced by a definition of metropolitan region very different from what we have today.

Although it now appears that major improvements in the commercial air industry during the 1940s and 1950s made it inevitable that larger airports far removed from the congestion of city centers would appear, such knowledge was not uniformly accepted by all city leaders during that same period. Those same leaders had helped subsidize the nation's commercial air industry by committing large amounts of money to airport development.

Eric Monkonnen has pointed out in his important book, *America Becomes*

Urban, that it is wrong to succumb to technological determinism when discussing the impact of transportation technologies on the city. Rather, Monkonnen observes, one should focus on how city leaders responded to different transportation technologies, for it is that response and accommodation that truly shape the city.[1] This essay not only shares the assumptions of Monkonnen's book, but goes further when discussing the relationship between commercial aviation and Dallas, proposing that in order to understand how Dallas civic leaders responded to aviation and airport development, one also needs to understand their assumptions about the nature of the metropolitan region. Their perspective put them at odds with a technology that some thought required a different way of looking at metropolitan regions. As a result, changing notions of metropolitan regionalism clearly help explain the differing responses to the idea of a Dallas–Fort Worth regional airport between 1940 and 1965.

Dallas in the 1920s experienced rapid change and growth. The city's population increased by 112,000 to 260,475 in 1930, and its physical size doubled from 23.4 square miles to 45.09 square miles.[2] Those occurrences did not just happen but reflected the accomplishments of aggressive civic leadership interested in promoting growth and development of Dallas. The city's decision to establish and nurture a municipal airport was part of that aggressive effort.

The city secured what would become Love Field when civic leaders under the leadership of the Chamber of Commerce leased land from farmers to secure a flying school for the U.S. Army Air Corps. After the war, when the army no longer needed a training school in Dallas and abandoned the field, the citizens who had leased the field used their option to buy and developed an airport for the city. During its early years, the airfield saw a number of barnstormers and even an occasional flying circus, but no scheduled air service. That changed after May 12, 1926, when the federal government initiated airmail service at Love Field. By this time, two things had become apparent. First, aviation held possibilities for future urban economic development. Second, the private investors of the Love Field Corporation had inadequate resources to undertake expansion of the city's only airfield. As a result, the Chamber of Commerce, guardians of the economic future, started lobbying for municipal acquisition of an airfield. Although Love Field seemed the logical choice for such a field, Dallas Chamber officials approached Fort Worth about developing a joint airport to secure more airmail flights and cut expenses for this new and untested municipal venture. Fort Worth declined, arguing that developing an airport in Dallas County would actually make it a Dallas airport.[3] Meanwhile, other Dallas leaders rejected the joint airport proposal and recommended a downtown airport built on the Trinity River bottoms adjacent to the central business district (CBD). Despite this, sentiment most favored the Love Field site.[4]

On December 17, Dallas voters passed a $400,000 general obligation bond for the purchase of Love Field to develop it as a municipal airport. The plan was viewed both as a way to strengthen the city's already enviable position as a distribution center and as a stimulant for attracting aircraft production to the city. One observer excitedly predicted that the purchase of Love Field was "the biggest thing that has happened for Dallas since the establishment of the headquarters of the 11th District Federal Reserve Banks." M. J. Norrell, general manager of the Dallas Chamber, concurred. According to Norrell, the purchase of the municipal field would go down in the city's history as "one of the turning points in [its] path of progress."[5] Airport development, then, was part of the larger booster strategy of building a bigger Dallas.

After Dallas citizens approved a new city charter in 1930, which replaced the city commission form of government with a more businesslike city manager/council form of government, the city made new commitments to improve Love Field. With the prodding of the Chamber of Commerce, the city proposed a new aviation bond election in 1931. Dallas voters approved the sale of bonds that allowed the city to pave airport runways, install a floodlight system for night flight, and purchase the most modern apparatus for control of air traffic. By 1934, the city manager reported that Love Field had been transformed from an unfenced, weed-covered field into a modern airport.[6]

Despite these improvements, officials had to hold another bond election in 1938 when the airport lost its top ranking from the federal government. The airport saw the number of takeoffs and landings triple between 1933 and 1937, but it still suffered because the newly developed DC-3 found Dallas runways inadequate. To solve that problem, local officials used the bond money to acquire additional land and to expand and improve the field's runways. In addition, local officials secured a $35,000 WPA labor grant to help build a new terminal.[7]

Except for the WPA grant, Dallas relied on tax money to build its airport. Until 1938, airport development was really a local concern. But after that date, federal funding and federal regulations would play a major role in shaping the airport and in resurrecting the regional airport issue. The Civil Aeronautics Act of 1938 not only proved an important milestone in this nation's aviation history, but also greatly influenced the relationship between cities and their federal government. First, the act called for a six-year national airport improvement program. Second, Congress for the first time authorized federal monies specifically for airport betterment. Finally, the act also expanded the federal government's regulatory power. It gave the Civilian Aeronautics Authority (later the Civilian Aeronautics Board) power to assign all new commercial air route service. This licensing power would have a definite impact on municipal airport development.[8]

The CAA's interest in developing a national airport plan once again raised the issue of a joint Dallas–Fort Worth venture. Responding to the CAA's request, Texas established an Aeronautics Advisory Committee to draw up a master plan of Texas airport development for submission to that federal agency. During the proceedings, the Texas State Aeronautics Advisory Committee discussed the possibility of recommending a joint airport for Dallas–Fort Worth but rejected it due to Fort Worth's strenuous protests. Dallas was interested in such an airport because it believed that the availability of federal aid would be greater for a joint undertaking. Dallas representative to the committee, D. H. Byrd, also thought a joint airport was "inevitable" because the airlines would demand it as a way of cutting down expenses due to costly stops in Dallas and Fort Worth. D. L. Johnson, chair of the Fort Worth Chamber of Commerce, challenged "the inevitability of it," arguing that the distance to a midway airport would be so great as to discourage its use. Indeed, Fort Worth supporters contended that the two cities were so far apart that "a common airport [would] serve neither to advantage." Fort Worth authorities also worried that any joint airport would in reality be a Dallas airport subsidized by Fort Worth. Fort Worth supporters were particularly concerned that any midway airport would probably be built in Dallas County and not in Fort Worth's Tarrant County. Such fears help explain why Fort Worth's member of the Advisory Committee argued that the thirty miles separating the two cities warranted separate airports. The committee's final report agreed and recommended that both Dallas and Fort Worth develop class 4 airports, the committee's highest-ranked airports.[9] Shortly after the meeting, Dallas lost some of its eagerness for a midway airport upon learning that the CAA would not immediately produce $1.8 million for a joint airfield.[10] Neither city was particularly enthusiastic about a joint airport, although the larger Dallas thought that a joint venture in what still appeared to be a risky urban undertaking might be feasible if heavily subsidized by the federal government.

Despite the Advisory Committee's recommendations, the CAA regional office continued to press officials from both cities to support a joint airport. To gain their attention, it refused to honor either city's request for federal funding of local airport improvement. Several airlines serving the area, including American and Braniff, joined with the CAA to advocate a regional airport because of the high costs and inefficiency of operating larger aircraft between the two cities.[11]

When Dallas and Fort Worth failed to cooperate with the CAA, the agency took another approach. It invited the small town of Arlington, located between Dallas and Fort Worth, to sponsor the proposed regional airport. American and Braniff airlines agreed to purchase the necessary land and turn it over to Arlington. Meanwhile the CAA agreed to construct an airfield with runways between 3,400 and 4,500 feet in length. Unwilling to see Arlington benefit alone from the government's generosity, Fort Worth and Dallas reentered the regional airport picture. In October of 1941, the three cities signed a tri-city pact to establish Midway Airport. According to the agreement, the airlines would purchase 1,000 acres of land and deed it to the three cities, which would in turn form a corporation to construct hangars, repair shops, and a terminal. The federal government allocated $490,000 for the actual runways and control tower.[12]

Despite the agreement, the cooperative airport never came to be. Personality conflicts between Dallas Mayor J. Woodall Rodgers and Fort Worth's leading citizen, newspaperman Amon Carter (a major stockholder in American Airlines), created a serious rift over the final location of the airport administration building. When plans for the building were altered from the preliminary site to one more advantageous to Fort Worth, Rodgers exploded. He charged that Carter's whining had convinced the government to change its plans and warned that such action marked the beginning of a "progressive steal" of the airport by Fort Worth. He angrily called off negotiations and rebuked the change of plans as a "monumental insult" to Dallas. Although some Dallas civic leaders thought the controversy over administration building location should not interfere with the development of Midway Airport, the city's leading newspaper concluded otherwise. According to the *Dallas Morning News*, "This [was] not a situation to be laughed off as being too childish to squabble over."[13]

Dallas withdrew from the airport agreement, but the CAA proceeded to build its runway for the army under the Landing Areas for National Defense Program and turned the field over to Arlington. Meanwhile, the city employed a planning expert to help it develop its own airport program. Dallas leaders feared that Love Field would soon be inadequate for the new four-engine aircraft coming into service. At the time of World War II, the airfield's longest runway was 3,730 feet while two other runways were under 3,000 feet. Wedged in between Bachman Lake on the north and neighborhoods, commercial districts, and busy streets on the south, east, and west, Love Field's status as the area's leading airport seemed doomed. A master plan developed for the city between 1943 and 1945 appeared to confirm this.

The city's decision to hire Harland Bartholomew and Associates in June of 1943 reflected its commitment to developing a comprehensive plan for metropolitan Dallas for the next twenty-five to thirty years. And that commitment reflected certain assumptions about the metropolitan area. Between 1920 and 1955, civic leaders not only in Dallas, but also in Cincinnati, New York, and elsewhere worked under the assumption that their metropolitan areas were real cultural units that affected how people behaved. Roderick McKenzie, author of *The Metropolitan Community* for the President's Research Committee on Social Trends in 1933, reflected this assumption in chapter 6, titled "Metropolitan Region as an Economic and Social Unit." He emphasized that the city and its metropolitan region were "more than an aggregation of peoples, or an agglomeration of buildings." He concluded that they constituted "an organization of actions, an economic and social organism." [14] Zane L. Miller has recently observed how the Chicago school sociologist defined the metropolis during this era as "a pluralistic cultural unit with its very own life." Moreover, according to Miller, assumptions at that time suggested that territorial communities "molded the desires, values, aspirations and personalities of [their] inhabitants." [15] This emphasis on place helps explain why Dallas and Fort Worth were somewhat reluctant to cooperate on a regional airport built beyond the defined metropolitan region of either city. It also explains the ferocity of the rivalry between the two cities, viewed as centers of discrete metropolitan regions competing for growth and dominance in north-central Texas. Finally, such an emphasis helps explain both the nature and the concerns of Bartholomew's plan.

Although the first of fifteen planning reports discussed the character of the city, most others focused on certain urban needs including street, park, and school development, as well as housing and zoning issues. All were to be coordinated into a "comprehensive plan or policy designed to regulate future growth and gradually to effect a well-balanced, unified community." Such a strategy was necessary because the "city [was] made up of many small parts, all interrelated and interdependent." But Mayor J. Woodall Rodgers believed that of all the parts to be examined, "airport improvements have No. 1 place on the master plan." [16]

In report number 6, titled "Transportation Facilities," Bartholomew laid out his comprehensive approach to airport development. According to the planner, one or two airports would not solve the city's future needs. Rather, the city required separate airports for commercial, military, air cargo, and private flight needs. As a result, he proposed a system of twenty-one airports for the city: one super-airport with 10,000-foot runways, a major airport, a secondary airport for military use, ten minor airports, and eight helicopter ports or small plane landing strips.

Bartholomew's airport plan reflected the general planning tendency of the day to plan physical development along functional lines. His plan also mirrored the country's optimistic sense of possibilities in private aviation and reminds us how difficult indeed it was to plan for such a new and rapidly changing technology. Bartholomew's concern with identifying and planning for the "real" Dallas metropolitan area also would affect the nature of his airport recommendations.

According to Bartholomew's studies, some two hundred square miles, corresponding with the city's natural watershed, could expect urban development as metropolitan Dallas during the next twenty-five years. "This future urban area," according to Bartholomew, "is physically, socially and economically a single unit" even though only one-fourth of it lay within the city limits. The plan, then, attempted to promote the development of the entire area "in a balanced and coherent manner."[17]

Bartholomew particularly worried about "a scattered, abnormally decentralized city" and its impact on metropolitan community. According to the planner, Dallas had two options. It could either allow the decentralization to continue and "give up the city" or "retain the present form of centralized city" by gradually stabilizing the present community and gradually rebuilding the obsolete districts.[18] It is in this context that Bartholomew's attitudes about a midway airport make sense. Bartholomew rejected the idea of developing a joint airport with Fort Worth some nineteen miles from the Dallas CBD because such an airport was not "practical . . . for intensive local commercial use. While there would be greater convenience and economy for transcontinental air lines to have a single airport for Dallas and Fort Worth passengers," Bartholomew feared the consequences for Dallas. "There would be much greater inconvenience, greater expense and waste of time for all passengers," he warned. "The losses capitalized over a period of years would be a staggering sum. No large city has its main passenger airline depot at such a great distance from the business district or from the center of population. It just does not fit into any comprehensive plans of airports to serve the future needs of Dallas."[19] Moreover, the airport site lay beyond the city's metropolitan area and would promote the very decentralization that Bartholomew feared. Such views reinforced leaders' prejudices against the joint project.[20]

In the same report, Bartholomew also questioned the utility of Love Field as the city's future airport. Although Love Field seemed destined to appear as an important part in the city's present aviation picture, the planner thought that the cost of making Love Field Dallas's future super-airport was "prohibitive, particularly in view of the much greater and far superior facilities that could be provided elsewhere for the same amount of money." As a result, Bartholomew recommended that only stage one of a three-stage plan for the expansion of Love Field be implemented to provide adequate air facilities while the city built a new airport. Once that was done, Love Field could be converted into a manufacturing, sales, and service airport.[21]

Bartholomew's plan recommended that the city acquire a site in southwest Dallas in the Lake June area for the future super-airport. That airport would cover 4,400 acres and include 10,000-foot runways and enough room to expand them to 15,000 feet. Such a massive size would allow Dallas to accommodate the largest air passenger and cargo ships imaginable. Located about twelve miles from the CBD and a twenty-minute drive from downtown, the site fell within what Bartholomew viewed as the natural metropolitan region.[22]

At the very time that Bartholomew prepared his plans, the U.S. Army decided to expand Love Field and make it a major base of operation. Near the end of 1943, the army agreed to spend more than $6 million over the next three years to improve Love Field for the Air Force Ferrying Command. The army proposed to expand the airport's runways 6,200 feet and make other improvements if the city agreed to secure the necessary land. Local officials accepted the army's conditions at the same time it approved the acquisition of the new airport site in southwest Dallas. Despite this commitment to developing an all-new "super-airport," the army's expenditures on Love Field, just seven and a half miles and a fifteen-minute drive from downtown Dallas, clearly revived interest in making Love the primary field after the war. According to the *Dallas Morning News,* the Lake June airport site would be held in reserve until 10,000-foot runways were needed. Meanwhile, Love Field remained as the city's main airport.[23]

About the same time Dallas employed Bartholomew and Associates to provide among other things airport planning advice, Fort Worth engaged the local engineering firm of Carter and Burgess to locate an airport site for that city. Since its current airport, Meacham Field, did not have runways long enough to accommodate four-engine planes, and since most thought that the airfield was located in an area inappropriate for expansion, the city instructed the engineers to locate a new airport site. They selected a site in the Hemphill section south of the city for the new airport. The engineers rejected the army's Midway Airfield because certain topographic features, as well as man-made obstructions such as nearby railroad tracks and highways, would make future expansion difficult. They also deemed Midway inappropriate because of its nineteen-mile distance from downtown.[24]

At the end of 1946, then, it appeared that North Texas would have two separate airports rather than the joint one pushed by the federal government and the airlines. Urban rivalry between two discrete metropolitan areas, as well as recommendations by expert planners, seemed the major reason for this development. Convenience and accessibility for each city took precedence over the convenience and efficiency for the airlines serving the two cities.

During the time Dallas and Fort Worth were planning their future airports, Congress passed the Federal Airport Act of 1946, which called for more federal airport planning. Indeed, the act mandated that the CAA develop a national airport plan to promote a system of public airports. Congress also authorized increased federal funding to develop that system. When federal officials examined the North Texas area they reaffirmed their belief that a joint regional airport seemed best.[25] Unable to secure Dallas's cooperation, federal officials convinced Fort Worth officials, whose city had lost ground to Dallas in the 1940s, to alter their airport plans and develop the now abandoned Midway Field as their municipal airport.[26]

Dallas leaders first reacted cautiously to this development, but became infuriated when the CAA released its national airport plan for 1948. That plan proposed Midway Field, rather than Love Field, as the area's primary airport. To see its hated rival, Fort Worth, gain the support of the federal government was too much for some Dallas leaders. They turned to their congressman, J. Frank Wilson, for help. When he failed to alter the plan, Dallas leaders took the CAA to court but lost again.[27]

As work finally started to convert the army's Midway Field into Fort Worth's commercial airport, renamed Carter Field, some Dallas leaders began to discuss the possibility of joining Fort Worth in its development of the old army airfield. In April of 1951, the city's most powerful civic group, the Dallas Citizens Council, voted to study such a possibility. Amon Carter of Fort Worth personally offered Dallas an opportunity to join in the airport's development, and one Dallas newspaper reported that "it is no secret that some of the most influential businessmen have changed their thinking about the 19 mile airport [Carter Field] and are ready to negotiate with Fort Worth on any basis." John W. Carpenter, president of the Dallas Chamber of Commerce, emerged as the most vocal supporter of the joint development of Carter Field. "If you disregard personalities and disregard politics," Carpenter observed, "you cannot fail to see that Dallas should support the Midway Airport as it originally did."[28]

Dallas-based Braniff Airlines President Thomas Braniff also urged Dallas leaders to cooperate with Fort Worth on a new airport. He feared that Love

Field would soon reach a saturation point and be simply unable to handle additional traffic safely. Stanley Marcus, president of the elegant Nieman-Marcus department store, also had doubts about Love Field's future. He agreed to chair a Chamber of Commerce committee on Love Field only after the Chamber convinced city council to engage an outside airport consultant to evaluate Love Field. Bond brokers also pressured council. They urged the city to secure outside expertise because the city planned to finance future airport development in part by selling revenue bonds to be paid back solely from airport-generated revenue.[29]

Council's decision in 1951 to employ James C. Buckley, terminal and transportation consultant from New York City, proved an important turning point in the history of Love Field. The forty-two-year-old Buckley had served as the director of airport development for the Port of New York Authority and played a major role in planning LaGuardia and Idlewild airports. He left that post in 1949 due to ill health and established a consulting firm on Fortieth Street. Dallas leaders employed him to evaluate the future of Love Field with the opening of Fort Worth's new airport scheduled to occur in about one year. Would Love Field survive the challenge of this new airport well enough to protect investments of both private and public monies? Buckley's investigation answered that question in the affirmative, claiming that a revamped Love Field would retain a minimum of 65 percent of the area's commercial aviation traffic. The reason was simple, according to Buckley. Since a large percentage of flyers were downtown businessmen, they would continue to rely on the convenience of Love Field long after Carter Field opened. In many ways Buckley echoed Bartholomew's opinion that the Dallas metropolitan area was a discrete region and deserved its own airport.[30] Buckley not only predicted Love Field's future, but he also underscored its present economic importance to the city. Love Field provided more than 3,600 jobs and a payroll of over $74.5 million. In addition, its passenger and freight cargo service played a critical role in promoting broader economic development. According to one account, Buckley's report surprised and informed Dallas business leaders who had not realized the impact of Love Field.[31]

Probably no single document did more to rally local support behind Love Field than Buckley's final March 1952 report. Before this time, many of the city's leading businessmen had viewed the Dallas–Fort Worth airport feud as primarily a conflict of personalities between the cities' leaders, most notably Fort Worth's Amon Carter and Dallas Mayor Woodall Rodgers. But Buckley's data emphasized the important role the municipal airport played in the city's economic life. As a result, local leaders were now willing to rally behind the expensive expansion of Love Field, something Buckley called for in his report.

Although Buckley urged cooperation with Fort Worth for long-term regional airport development, his report gave a ringing endorsement to Love Field and would shape the airport policy for the next ten years. Buckley not only recommended that Love Field remain the city's primary airport, but argued that Dallas should oppose any schedule pattern filed by the airlines for the Dallas–Fort Worth area that failed to provide that all local airline service stop at Love Field. He also suggested that Dallas leaders do everything possible to encourage additional air service for their city.[32]

Responding to this last suggestion, the Dallas Chamber of Commerce employed Buckley to review the present and prospective air service requirements and identify significant gaps in the city's service. Buckley's report proposed new and improved service to seventy-five communities based on market and economic ties between Dallas and those communities. The Chamber used his rich data and sound arguments in its quest to secure more air service from the Civil Aeronautics Board.[33]

Buckley's two reports, one to the city council and the other to the Chamber of Commerce, helped launch an unprecedented Love Field promotion. City government pushed for the physical development of the airport while the Chamber of Commerce encouraged additional air service. Whatever the focus, both worked under the shared assumption that Love Field was a valuable asset to greater Dallas, and its future development promised even greater riches for the city.

One of the most important manifestations of the new commitment to Love Field came with the $10 million airport bond proposal in 1953. Airport improvements to the tune of $5 million would be undertaken with the sale of revenue bonds, issued solely against anticipated airport income, but local officials needed other money for their ambitious program of runway expansion and a new terminal building. Never before had Dallas voters been asked to approve such a hefty sum for airport development. To insure that the measure carried, supporters spent more than $50,000 in the campaign, claiming that airport improvement was one of the most important issues on which the public had ever voted.[34] Robert L. Thornton, possibly the most powerful leader in Dallas, warned that his city was at a crossroads. "We must go forward," he cautioned, "or be like some of the towns that the railroads passed up. If we don't follow the word of the experts who say we must expand Love Field, what word are we to follow?" Disparate groups such as the Dallas Building Trades Council, the Negro Chamber of Commerce, the Dallas Home Builders Association, and the Oak Cliff Chamber of Commerce supported the bond.[35]

Others in Dallas, however, opposed the expansion of Love Field. More than seven hundred African Americans living in nearby Elm Thicket jammed the North Temple Baptist Church on January 19 and urged defeat of the airport bond. Wondering where those who would lose their homes to airport development would go, optometrist J. O. Chisum warned that additional airport expansion would push blacks into "unfriendly [white] communities." Another speaker at the meeting feared for the safety of those remaining in nearby Love Field neighborhoods. Citing the Doolittle Report on air safety, published in 1952 after three separate airplane crashes had occurred in a six-week period near Newark Airport, he concluded that Love Field failed to meet the Doolittle Commission's recommended safe zone and runway length standards. Members of this meeting formed the Dallas Home Owners Protective Association and campaigned hard against the bond issue.

A second group concerned about the safety hazards in the congested Love Field area also opposed the bond issue and the expansion of the airport. The Air Safety Committee warned that runway extension would pose a direct threat to nearby schools. It emphasized that a new airport could be built on the city's fringes for about the same cost as that of Love Field expansion. Despite the opposition of these groups, more than 19,000 Dallas voters agreed that Love Field would bring economic benefits to greater Dallas and voted for the bond issue, giving it a 4,000-vote victory.[36]

Eleven days later, Buckley completed a detailed master plan for Love Field's future development. This revision of an earlier plan developed during Bartholomew's tenure called for airport development in three stages with the ultimate extension of the northwest/southeast runway from its current length of 6,200 feet to 8,500 feet. It also called for the completion of a parallel northwest/southeast runway and the erection of long hangars, an airline maintenance facility, and a new terminal building. These improvements, according to Buckley, would help Love Field maintain its place as the area's primary airport. City council accepted the master plan revision and appointed a special aviation committee to carry it out. According to one observer, Love Field was Dallas's "number one municipal project." The 231 resolutions relating to airport development passed by city council in 1954 seemed to confirm that observation.[37]

The decision to expand Love Field came at a time when Dallas leaders felt the effects of Fort Worth's new airport, Carter Field. After Carter Field opened on April 23, 1953, Fort Worth leaders undertook an aggressive campaign to secure more commercial air traffic for the airport. American Airlines moved nine of its flights from Love Field to Carter Field, much to the joy of its largest stockholder, Amon Carter. Tom Braniff also toyed with the idea of moving much of his operation to Carter Field. He decided to stay at Love after Dallas officials promised to help him finance new facilities there.[38]

Fort Worth not only attempted to lure established traffic from Love Field, but it constantly intervened in Dallas's attempts to secure new flights for its field. Whenever an airline applied for new routes into the North Texas region, Fort Worth pushed the CAB to make Carter Field the terminus for the entire region. Fort Worth's first success sent shock waves throughout Dallas. When Central Airlines requested permission to fly into the Dallas–Fort Worth area from Oklahoma, Fort Worth succeeded in having the CAB designate it the sole terminus. The CAB's action reflected its commitment to promoting a regional airport for the Dallas–Fort Worth area. And Dallas officials understood the awful significance of that act to Love Field since the CAB's power to designate airline travel to a single airport in a region could thwart any future development of Love Field traffic and make Carter Field, without Dallas input, the de facto regional field of the area.[39]

Dallas officials challenged the CAB's action in federal court and lost. When the CAB ruled against Dallas several months later in another air service case, Dallas civic leaders became infuriated. Chamber of Commerce President W. W. Overton condemned the CAB's regional strategy and argued that the board's action showed a "scandalous disregard of [its] obligation" to build a sound air transportation system to serve Dallas traffic.

Although Dallas leaders succeeded in securing some additional routes for Love Field before the year ended, new controversy soon appeared. Under the prodding of CAB chair Char Guerny, Fort Worth offered on November 14, 1954, to sell a half-interest in Carter Field to Dallas. Angered at Guerney's actions, Dallas Chamber of Commerce President Jerome K. Crossman accused the CAB head of injecting himself "extra-legally into the affairs of [the] community." Several days later, Buckley advised the city against accepting the offer because of the negative impact the move would have on Love Field's employment and payroll. After consulting with council, Mayor Rodgers heeded Buckley's advice and formally rejected the proposal, despite threats from Assistant Secretary of Commerce John R. Allison that he would push Carter as the transcontinental route terminal for both Dallas and Fort Worth. Dallas's unwillingness to cooperate with Fort Worth had already led the CAB to announce that it would approve any airline's request to move from Love Field to Carter Field. Dallas, which had earlier tolerated federal intervention in the airport business, now bristled at action that seemingly favored its rival and posed an economic threat to the city's welfare. The conflicting goals of federal policy, which emphasized an efficient, safe, and cost-efficient national air system, and Dallas policy, which stressed economic development for the city and dominance over the region, created great antipathy between local leaders and federal aviation officials.

Despite the federal government's firm commitment to regional airport development, it appeared to ease its campaign in the mid-1950s. Not only did the CAB start granting Love Field more flights, but in 1956 the FAA even allocated $375,000 to Love Field for airport expansion. In addition, it reclassified Love Field from continental to intercontinental status. Softening of the federal government's campaign to force Dallas into a regional airport agreement may have contributed to the failure of Willow Run Regional Airport developed for greater Detroit. The city abandoned that airport, which was more than thirty miles from Detroit's downtown. Baltimore's Friendship Airport, midway between that city and Washington, also had not been very successful. Fort Worth's Carter Field seemed destined for the same fate as its number of passengers and planes steadily declined during the mid-fifties.[40] These developments, as well as the dedication of Love Field's new $7.5 million terminal in 1957, led the *Dallas Morning News* to celebrate the Dallas airport's apparently secure status in an editorial titled "Love Field Triumphs After Long Struggle."[41]

Emergence of noisy commercial jet travel in the late 1950s, and continued congestion of Love Field, however, still made that field vulnerable and kept the regional airport idea alive.[42] Not only did jets intensify the area's noise problem, but they required longer runways. The FAA's airport plan for 1959 called for officials to develop parallel runways of 9,200 feet for Love Field, 700 feet longer than Buckley's master plan had suggested. That and other signals from the federal government worried local officials, who responded by initiating a $9 million expansion program for the airport in 1959 to complete the third and final phase of the 1954 Love Field master plan. Plans met a snag in November when the FAA rejected a Dallas request for $838,000. It refused the request, in part, because Love Field plans had not provided an adequate clear zone (safety zone at the end of the runway), nor was the proposed runway long enough to satisfy FAA safety standards.[43]

Adverse federal action was not the only barrier to successful completion of airport expansion. A lawsuit by forty-three Love Field area homeowners attempting to halt runway expansion plans delayed construction for nearly two years and helped refocus attention on the area's need for a larger, more regionally based airport.[44]

During the new controversy, which saw increased pressure on Dallas to cooperate with Fort Worth to make Carter Field a truly regional airport, Dallas officials again turned to Buckley for help. On April 25, 1960, city council asked the airport consultant to reevaluate the master plan and airport. It specifically asked him "to determine the extent, if any, to which Dallas's airline requirements are likely to grow beyond the capability of Love Field; and if such appears likely, to suggest possible methods for meeting such excess requirements."[45]

His findings reaffirmed Love Field's adequacy for "the foreseeable future and certainly over the next 20 to 25 years, insofar as runway length, capacity and usability be concerned." Even if Love Field's "excellent location and excellent facilities" brought it to an earlier saturation point, Buckley thought the city could still expand the airport rather than join with Fort Worth to develop Carter Field.[46] Armed with his recommendation, the city continued to disregard federal officials' requests for Dallas involvement with Carter Field.

The FAA proved just as stubborn. It refused to funnel additional monies into Love Field and made Carter Field the North Texas recipient of federal funds. Even more disconcerting for Dallas officials was a new joint FAA–CAB policy announcement in 1961 that expressly called for the development of regional airports.[47]

Matters only worsened when FAA head Najeeb E. Halaby, who incidentally had been born in Dallas, testified before a Senate subcommittee in 1962 that the FAA would "not put another nickel in Love Field." The FAA chief warned that new appropriations for Love Field would be a waste of money because of congestion and because of the field's inability to meet the demands of future aircraft. He chastised the city for not developing a regional airport with Fort Worth and suggested that Dallas's commitment to Love Field resulted from nothing more than a "pure, unadulterated case of childish civic pride."[48]

To city fathers justifiably proud of their airport built mostly with Dallas money, the remarks seemed a declaration of war. Dallas leaders, interested in maintaining an airport that would be most accessible to downtown business people, and that would promote the growth of metropolitan Dallas over that of nearby metropolitan rivals, had followed Buckley's expert advice and developed Love Field as a super-airport. Indeed, local officials had invested more than $32 million in the airport since 1928. Now federal officials, viewing the airport from a national perspective, threatened the field's existence, as well as the city's investment, by demanding that Dallas officials cooperate in Carter Field development. The federal government's demands struck Dallas leaders as unfair since local tax money had played such a major role in Love Field, while federal dollars had significantly financed Carter Field. According to one estimate, Fort Worth had received $4.70 in federal airport aid for every passenger using Carter Field while Dallas received about 43 cents in federal aid per passenger at Love Field.

The federal government's demands that Dallas work with Fort Worth to develop Carter Field into a true regional airport also seemed inappropriate since Fort Worth's own engineers had questioned whether the geography around Carter Field would ever allow it to become the massive airport that some federal officials wanted it to become. Finally, Dallas officials understood that an airport located such a distance from Dallas would encourage even more population deconcentration. The city already was experiencing the economic impact of suburban migration and feared the consequences of an airport located far beyond the city's corporate limits. This context helps explain why Mayor Earle Cabell promised to fight the federal government over the airport "with every weapon at [his] command." Others echoed the mayor's resentment at the federal government. Chamber of Commerce Aviation Committee chair H. L. Nichols complained of the "deliberate and massive attack upon Dallas and its airport . . . by two agencies of the federal government." He particularly protested what he saw as a "misuse of Federal Power in an effort to dictate to Dallas a course of action contrary to Dallas's own interests."[49]

The CAB hearing over whether Dallas–Fort Worth should have a regional airport, and if so whether it would be Carter or Love Field, started on July 7, 1963. Examiner Ross Newman began the hearing in Arlington, Texas, midway between the rival cities, and finished it on September 20 in Washington. During those weeks, FAA officials testified against Love Field, emphasizing the noise problem at the Dallas airport. Even more telling, the FAA claimed that Love Field had neither adequate runways nor the capability for instrumentation to safeguard flights the same way Carter Field could. One FAA official listed seven factors that would keep Love Field from being an acceptable regional airport.[50]

Dallas spokesmen countered that their airport met FAA safety requirements. They also emphasized how Love Field's convenience saved Dallas travelers time and money. Moreover, they reminded Newman that Love Field was a cornerstone of the economy. And when James Buckley met with the CAB, he testified that Dallas and Fort Worth were "*separate and different types of communities and economically incompatible in their service air requirements*" (italics added).[51] That is, the consultant maintained that Dallas and Fort Worth served different "metropolitan regions," discrete places with little in common. But as we have seen, the CAB discounted this organic interpretation of the metropolis that had been promoted by both Bartholomew and Buckley and embraced one that included both Dallas and Fort Worth in a single metropolitan region.

So angered were they at the possibility of having Carter designated the regional airport that Dallas officials threatened to have absolutely no involvement in that airport even if the CAB gave it regional status. The Dallas threat to refuse any financial help for Carter and to close Love Field down if ruled against put the CAB in an awkward situation since it was doubtful that Carter Field could truly be an adequate regional airport without Dallas's help.

These new threats may help explain why examiner Ross Newman decided not to designate a regional airport to serve Dallas–Fort Worth. But that decision was short-lived for on June 12, 1964, the CAB announced it would review Newman's decision. After that review, the CAB announced on September 30, 1964, that both cities would be served by a single facility. It gave the two cities 180 days to find a location, threatening to locate the site itself if the cities failed.[52]

Although the decision outraged some Dallas leaders, others, weary from battling the federal government, tired of negative publicity, and concerned that Houston might capture Dallas's long lead in aviation if something was not done, acceded to the pronouncement and started secret negotiations with Fort Worth. Unable to reach an agreement as to the proposed airport site in the time allocated by the CAB, the cities were granted an extension by a federal agency strongly wanting a local solution. The CAB's patience paid off when on May 31, 1965, both cities' councils ratified an agreement to build a regional airport. In reporting the new willingness of Dallas to cooperate, the *Dallas Morning News* observed how this was a "great departure from the Dallas thinking of just a few years ago," noting that such action back then would have amounted to "political suicide." The paper attempted to explain the new cooperative spirit by noting a changing leadership in both Dallas and Fort Worth to individuals not connected with old arguments and animosities.[53] But the new willingness of Dallas officials to obey a federal mandate rather than to challenge it may also have reflected a change in thinking about metropolitan region. By their earlier actions, they defined the region as a real cultural, social, and economic unit that influenced individual behavior. Their abrupt about-face suggests that they now treated it as a neutral setting for individuals to pursue their own ambitions. Indeed, Zane Miller has argued that the 1960s saw the decline of a social taxonomy that defined the metropolis and region as organic and as basic units of society. Instead, he claims "people took autonomous individuals as basic units of society."[54] Evidence for this change can be found in the efforts to find the proper term to define the Dallas–Fort Worth area. The terms most associated with an organic definition—metropolis and metropolitan region—were infrequently used. Others, such as the Dallas–Fort Worth–Denton Megapolis and Metroland, were employed for a while. Finally, the term "metroplex" was coined for the area. Harvee Chapman of the Troy-Locke advertising firm is usually credited with its invention. Chapman's firm was hired in 1971 by the North Texas Commission to sell the north-central Texas region to the country. The North Texas Commission had been established by local officials and civic leaders to promote the area after construction of the regional airport got under way. Today, then, the name most closely associated with the Dallas–Fort Worth area was part of a market strategy, to identify a service area tied together by highways and communication facilities, anchored not by a city but by an airport.[55] Unlike the metropolitan community Roderick McKenzie wrote about in the 1930s, the new unit made no pretense about organicism. Rather it was sold as a setting for individuals to pursue individual desires and ambitions.

The Love Field controversy provides insights to a number of issues including the persistence of urban rivalry and urban-federal relationships. The controversy underscores the tensions between planning from a national context and planning from a local context. Not only were the national planning goals—promotion of a safe and efficient national airway system—different from the Dallas goals—promotion of an economically prosperous city—but there appeared differences as regards what constituted the appropriate metropolitan region for which to plan. Dallas leaders and planners viewed greater Dallas, not greater Dallas–Fort Worth, as a definable, cultural, social, and economic unit. From the Dallas perspective, then, sophisticated Big D and cowtown Fort Worth were clearly discrete places. From the federal perspective, however, both North Texas cities shared broadly common social, cultural, and economic traits. Indeed, from the national perspective, Dallas and Fort Worth seemed to share more in common with each other than with the rest of the country.

Finally, this essay suggests that not only changing aircraft technologies but also changing definitions about the nature of metropolitan regions may help explain why Dallas leaders finally abandoned Love Field as their main airport, for a regional airport eventually located midway between Dallas and Fort Worth, just north of the old Carter Field. When Dallas leaders viewed their metropolitan community as organic and struggled with the discrete metropolitan community of Forth Worth for dominance of North Texas, they emphasized competition and noncooperation with their rival. By the 1960s, the view of metropolitan regionalism as an organic unit gave way to a new, nonorganic economic definition. And that allowed Dallas leaders to see the Dallas–Fort Worth area through different eyes. Indeed, it permitted Dallas leaders to put away their chauvinism and reach a regional airport agreement with Fort Worth that would bring unprecedented growth to north-central Texas.

Introduction to Chapter 10

In his essay, Alan I Marcus addresses a knotty problem in contemporary America: government regulation of the environment, drugs, cosmetics, and food products, in this case diethylstilbestrol (DES), a growth hormone used to promote growth in cattle. DES is a naturally occurring growth hormone that, when given to feedlot cattle raised for market, effected, on average, superior growth in less time and with less feed than in cattle not so fed. Indeed, the profits were considerable. At the same time, some critics charged that DES was dangerous to human health because the drug was associated with cancer in laboratory rats.

Yet according to Marcus the problem of regulation became a problem in expertise and its legitimacy. In the later nineteenth century, Marcus notes, scientists, engineers, doctors, and other professionals were able to present themselves as disinterested, objective, and impartial "experts" who imparted expertise, meaning knowledge derived truthfully and without partisanship toward one set of interests or another. What the debate over DES shows, Marcus insists, is that the very conception of expertise as the collective judgment and wisdom of disinterested authorities had little or no currency in American society and culture. In a word, technical knowledge stopped being the province of experts. The conception itself lost all meaning. For in an age of individualism and individuation, which that since the 1950s certainly has been, each individual assumes his or her autonomy in all matters—including the expertise needed for all manner of judgments. Marcus perceives this phenomenon taking place among those groups and constituencies whose members had previously been receptive to expert opinion and to those who had presented themselves as the said Delphic experts.

Marcus thus sees the DES controversy as symptomatic of tendencies in the larger culture itself. The debate over DES involved far more than technical issues surrounding the cancer-producing potentiality of beef fed with DES. At bottom, Marcus argues, the objectivity and authority of experts is no longer taken seriously, for in the DES controversy each side had its own experts, and the battle was judged to be won when public opinion, not the opinions of the experts, weighed more heavily with one side than with the other. Each side fought to manipulate the mass media in its favor, which suggests what regulation and, for that matter, expertise have come to mean in our time. Thus technical knowledge would seem to have come full circle, as suggested in the essays in this book, from knowledge of how to be a successful individual to knowledge of how to be a politically effective one. Imbedded within technical knowledge, then, are not immutable and unchanging truths about the world external to us, but, instead, larger cultural notions that shape and inform the ways in which contemporaries in a particular age interpret the world around them.

Unanticipated Aftertaste: Cancer, the Role of Science, and the Question of DES Beef in Late Twentieth-Century American Culture

10

On February 18, 1954, Iowa State College announced that Wise Burroughs of the Animal Husbandry Department had filed a patent application for a feed additive that speeded cattle growth by more than 10 percent. That substance was the popular synthetic estrogen diethylstilbestrol, commonly known as DES or stilbestrol. When as little as ten milligrams of this artificial female hormone was mixed daily in a beef animal's rations, that animal achieved market weight roughly thirty-five days sooner and consumed about five hundred pounds less feed. In addition to yielding beef more quickly and at a reduced unit cost, DES produced meat lower in fat and higher in protein.[1]

Mrs. Robert Swenson, a resident of River Falls, Wisconsin, injected a somber note when the next day she learned about Burroughs's discovery in her local paper and wrote to him to relate a story she had read in the "Science Report Tells" section of the February 1954 issue of the *National Police Gazette*. That article maintained that "75% of the rats tested" with DES got cancer, and Mrs. Swenson demanded to know "why is our government letting . . . manufacturers manufacture the drug." She asked incredulously if the government would "let the public suffer" and wondered whether "our public officials and educators are not aware of this harmful drug?"[2]

Burroughs sent a reassuring letter to Mrs. Swenson, but he surely did not anticipate the discord his patent eventually would reveal.[3] Although complaints not dissimilar to those uttered by Mrs. Swenson began to be issued sporadically as soon as the Food and Drug Administration licensed stilbestrol as a cattle growth promoter in late 1954, they became commonplace about a decade and a half later. Books bearing titles such as *The Poisons in Your Food, The Chemical Feast,* and *The Politics of Cancer* and hammering home the DES in beef threat abounded. Newspaper headlines alerted the public to "cancer seeds in meat," asked "who regulates the regulators," and contended that Americans were in the grip of a "panic over food additives."[4] By 1971, an estimated 95 percent of American beef cattle daily received DES. Also in that year, the Natural Resources Defense Council, in conjunction with the Environmental Defense Fund, National Welfare Rights Organization, and the Federation of Homemakers, filed a lawsuit to force the FDA to outlaw DES as a cattle feed additive. The Center for the Study of Responsive Law and the Center for Science in the Public Interest, both Nader organization entities, mobilized against DES in meat as did such groups as Consumer Action Now, Inc. and Drugs Out of Meat, otherwise known as DOOM. DOOM submitted a petition with more than twenty-five thousand signatures to the U.S. Senate in 1972 demanding legislation banning stilbestrol as a beef cattle growth promotant. Its sentiments typified those of many concerned groups. DOOM asked why "cancer continues to strike more often and much sooner than ever before" and found the "answer . . . in the food we are ingesting." It was "loaded with carcinogens," especially "the synthetic female sex hormone." What resulted besides cancer was "millions of dollars to the pharmaceutical industry, and to the cattle and sheep growers," a situation that led DOOM to ponder whether "life [is] that cheap in the United States, compared to profits of individual companies."[5]

Certainly the veracity or lack of veracity of complaints about DES-treated beef were significant. The food might sicken, even kill citizens. But the nature of the complaints and the act of complaining itself, phenomena much less likely to be explored, serve as focus here. They too are portentous, perhaps even more so than the DES danger issue. These broad-based sentiments reflect a new way of analyzing and defining an extant situation, a new way of thinking and consequently a new way of attacking "public" problems. A segment of American society had voiced its lack of confidence in an official government agency, the FDA, established presumably to protect the public interest. And although that may not have been so unusual, the sphere in which this dissatisfaction arose was much less common. A scientific agency's determinations had been disputed by a substantial portion of the public. Many in American society maintained in effect that the matter of safety of DES in beef, a question of the type long identified as within the province of science and expertise, was best adjudicated by nonscientists and nonexperts. That constituted direct repudiation of the time-honored notion, institutionalized in the United States during the late nineteenth and early twentieth centuries, that science and by extension those who pursued science—scientists—were virtually unassailable. Science had appeared nonpartisan, infallible, its findings reproducible and verifiable and not subject to debate. In the pursuit of science, scientists had become cloaked in science's mantle and long seemed beyond reproach. Science's vaunted ability to permit corroboration had placed a premium on consensus among its practitioners. It had established a de facto template for scientists, deviations from which branded individuals as either sloppy or quacks; vituperative public disagreements among scientists had not been tolerable and, ideally, not possible.[6]

This formulation stands in stark contrast to debates over DES beef. The case of DES was not unique, moreover. During this period in American history, nitrates, nitrites, nitrosomines, cyclamates, saccharin, red dye number 2, antibiotics in animal feed, and numerous other chemical substances underwent public scrutiny and professional debate. A veritable alphabet soup of regulatory agencies emerged during these years; the Environmental Protection Agency (EPA), Occupational Safety and Health Administration (OSHA), and National Institute for Occupational Safety and Health (NIOSH) were the best known of the nearly twenty such instrumentalities created in the decade after 1968.[7]

But it is the very typicality of the DES debates, a typicalness only engendered during these years, that renders them useful. They work as tools through which to gain understanding of the threads underlying and unifying American civilization in the 1960s, 1970s, and 1980s. Of critical importance was the fact that both opponents and proponents of DES as a cattle feed additive rejected the notion that governmental regulation was a scientific process, pursued by disinterested experts. The various tactics employed and arguments presented were symptomatic of a decline in faith in the uniformity, moral certainty, and nonpartisanship of experts generally and demonstrated the factionalization or, more properly, the individuation of American policy making, both scientific and otherwise. In sum, an examination of the DES controversy helps reveal assumptions new to the later twentieth century about scientists, scientific knowledge, and scientific inquiry; the relationship of government to the public; and methods of public adjudication. These in turn help us begin to recognize the individuation of America as the defining social element of the last half of the twentieth century.

Almost from the original synthesis of DES in 1938 scientists made crucial assumptions about the drug's properties. These assumptions rarely underwent sustained, in-depth scrutiny until the late 1960s. As early as 1938, studies on cancer-susceptible laboratory rodents demonstrated a statistically valid relationship between high doses of DES over extended periods and increased incidence of cancer. But scientists understood DES's carcinogenic properties as a by-product of the drug's estrogenic activity; they knew that each estrogen when ingested for long periods in doses large enough to produce anatomical or physiological changes was carcinogenic. Moreover, they recognized estrogens as naturally occurring in some plants as well as in all animals, an essential part of life.[8]

The ubiquity of estrogens in nature set the cancer question squarely on drug dosage. As long as stilbestrol was present in quantities so small as to not produce anatomical or physiological changes, it seemed harmless; it was natural. And with that idea in mind, Burroughs and others at Iowa State and Eli Lilly and Company refined in 1954 the highly sensitive, very accurate immature mouse uterine test, which took weeks to run. These investigators determined that the uteri of immature female mice would gain weight if the animals' diet contained more than about two parts per billion of diethylstilbestrol, a dosage grossly beneath the level necessary to produce similar changes in humans. To "prove" DES as a cattle growth promoter safe for public consumption, test mice were fed only DES beef. The repeated negative biological assays—the lack of detectable estrogenic activity—indicated less than two parts per billion of the drug in the meat and supposedly provided Americans a large margin of safety from any potential cancer hazard.[9]

During the following decades the Association of Official Analytical Chemists (AOAC), a collection of several thousand government-affiliated chemists who took their mandate as setting official methods of analysis for regulatory matters, grappled unsuccessfully with the problem of creating a quicker, cheaper, reliable test. The group's continual labors to devise a DES assay more suitable for regulatory work and persisting inability to achieve that end might have seemed moot, however, in wake of the Delaney Cancer Clause's passage. This 1958 act had banned substances from the food supply if "found to induce cancer [in] . . . man or animal." It had not outlawed all DES-laced animal feeds, though. Those firms previously licensed to manufacture feed containing the drug were grandfathered exceptions to the act, but standing licenses could be neither modified nor transferred, nor could new licenses be approved. Subsequent creation of near monopolistic conditions and high prices for DES animal feed premixes led Congress in 1962 to pass what was commonly known as the DES Amendment to the Delaney Cancer Clause. It specified that the Delaney clause did not apply to ingredients "of feed for animals which are raised for food production" when two main conditions were satisfied. First, the Secretary of Health, Education and Welfare, the department in which the FDA then resided, needed to set "conditions of use . . . reasonably certain to be followed in practice." Second, the secretary was required to stipulate that according to these conditions the additive "will not adversely affect the animals for which such feed is intended," nor would any residue of the additive be "found (by methods of examination prescribed or approved by the Secretary by regulations) in any edible portion of such animal after slaughter."[10]

The secretary relinquished his prerogative to the FDA commissioner as was custom, and as always the commissioner deferred to the AOAC for the official method. Since the AOAC was unable to devise a workable substitute, the commissioner fell back on best scientific practice and quickly certified the mouse uterine bioassay, the only reasonably verifiable method. He then turned to a study indicating that all DES left an animal body within twelve hours of ingestion and, to be extra cautious, mandated that farmers withdraw their cattle from DES-enhanced feed at least forty-eight hours prior to slaughter. A period of reduced public interest in DES-fed beef followed the commissioner's pronouncements. A joint National Academy of Sciences–National Research Council symposium on drugs in animal feeds did nothing to disrupt the calm. This prestigious 1967 symposium considered among other questions animal drug residues in meat consumed by humans. In the case of DES, the symposium reaffirmed that residues tiny enough so that no estrogenic activity was produced could not affect human health.[11]

DES as a cattle growth promotant had received en passant the imprimatur of America's foremost scientific agency, the National Academy of Sciences. In 1969, Elanco, the Eli Lilly subsidiary that manufactured DES feed premixes, capitalized on that assessment to introduce a more active, more potent form of DES.[12] Also in the same year, the company petitioned the FDA to permit doubling the daily amount of stilbestrol fed to cattle. As the FDA mulled over Elanco's request, the Consumer and Marketing Service of the United States Department of Agriculture—the entity charged with testing for DES in meat—found the drug in quantities slightly in excess of two parts per billion in a small percentage of slaughtered cattle. These USDA tests involved only a few hundred cattle with but a handful of positives, about one-half percent. No DES was recorded in the animal's muscle tissue; it was found only in the liver, the organ that metabolizes diethylstilbestrol into conjugates. The USDA's findings, coupled with the granting of Elanco's petition in 1970, heightened public concern, and the department later that year and in early 1971 responded to the clamor. USDA analysts began to experiment with a new speculative assay, gas-liquid chromatography (GLC). This assay required considerable skill to administer but was much faster and cheaper to perform than the mouse uterine method. GLC also was potentially much more sensitive—to less than one part per billion—but the USDA analysts had difficulty duplicating its DES results, and the AOAC had refused to sanctify its use.[13]

The department's disappointment that the new assay failed to achieve the reliability of the mouse uterine test was overshadowed by publication of Arthur Herbst's DES daughters study in April 1971. Harvard Medical School's Herbst had detected in seven women aged fifteen to twenty-two a form of vaginal cancer rarely found in the young and had set out to determine the outbreak's epidemiology. Aware of the almost indiscriminate use since the late 1940s of DES to prevent miscarriages and paradoxically as a morning-after pill to induce spontaneous abortions, Herbst learned that these young women's mothers probably had been given massive doses of DES (up to 125 milligrams daily) during pregnancy. He hypothesized that the drug had been transmitted transplacentally and suspected a causal relationship between diethylstilbestrol and the daughters' tumors.[14]

Reports of Herbst's work unleashed demands that the FDA prohibit DES use during pregnancy. They also produced cries that the agency ban it immediately from cattle feed. The United Nations Food and Agricultural Organization called for a worldwide proscription against DES, and several nations without large feeder cattle industries removed the drug from beef production. While the FDA did issue circulars to physicians stating that the use of DES was contraindicated during pregnancy, its move against the drug as a cattle growth promoter was less dramatic. The FDA merely lengthened the withdrawal period from two to seven days and required cattlemen to sign a certificate testifying that animals set for slaughter had been withdrawn from DES-added feed for the stipulated time.[15]

This reconsideration of DES in animal food also exposed a nasty dispute among FDA scientists. Assumptions about the nature of carcinogenesis were central. A handful of FDA investigators put forth the view that "a carcinogen is a carcinogen at any level," including the "one-molecule level," and repudiated traditional notions such as a "no effects level" and a "threshold level" as wrong and pernicious. They attacked the "safety factor" concept—the long-standing policy by which the smallest dosage of a drug found to cause cancer in laboratory animals was a hundred or a thousand times greater than the maximum permitted in human food—as inadequate and argued that ideas such as "virtual safety" ought to replace that type of thinking. Asserting that they had reached "a sound decision based on strictly scientific considerations," they defined virtual safety as the dosage at which only one out of 100,000,000 susceptible animals might get cancer, which they extrapolated from a log dosage/response curve, and relied on a statistical determination to arrive at what they contended was a 99 percent confidence rate. To achieve virtual safety in the case of DES would have necessitated that residue levels be more than 20,000 times below two parts per billion. Effective regulation at that level would have required a reliable, convenient assay sensitive to parts per one hundredth trillion, clearly not possible in the early 1970s.[16]

The vast majority of FDA scientists disagreed vehemently with the virtual safety contingent. They opposed that concept as "an emperical [sic], ultraconservative model which in turn is based on numerous assumptions," each of which was "not consistent with the judgments of experienced toxicologists." Noting that "there is no example of any chemical or toxin which is known to fit" the no threshold dose theory, they dismissed it as utterly speculative, not the product of science. "Estimates which are unreasonably small, as compared to the judgments based on generations of accumulated experience," they huffed, "must represent a paradox where 'must be' according to mathematical model simply cannot be according to practical experience." [17]

The status quo remained in effect at the FDA, but it was not the only agency to feel the pressure. By April 1972 the USDA had adopted the highly sensitive but erratic GLC assay as that agency redoubled its sampling for DES in edible beef. In July the USDA announced the results of its tests. About 3.5 percent of the several thousand samples of meat assayed in this latest round of USDA investigations tested positive for DES in any quantity—most in less than two parts per billion—and then only in livers or in a few cases kidneys, the site in which the drug and its metabolites are eliminated. Acknowledgment of this apparent seven-fold increase in stilbestrol-contaminated carcasses exacerbated tensions. But by then, the FDA had already initiated the laborious legal process required to remove DES as a cattle feed additive. Despite public claims to the contrary, the bureau could not ban DES by edict; it lacked the circumstances to invoke either the Delaney clause or the imminent hazard section of the Food, Drug and Cosmetic Act. The former did not apply because the approved method, the mouse uterine assay, turned up virtually no instances of DES-contaminated beef—the one-half percent was attributed to the carelessness of or disregard for regulations by renegade cattlemen—while the more sensitive USDA test was unreliable and therefore incapable of receiving scientific or the HEW secretary's sanction. The FDA could not claim the latter because despite the virtual safety crowd's pleas, the overwhelming majority of scientists continued to assume that DES in quantities beneath that required to stimulate anatomical or physiological changes in mice was harmless in humans. No responsible scientist in the government, including the vocal head of the National Cancer Institute, would categorically state that DES as then found in the meat supply definitely posed an imminent threat to human health. And without that testimony, the FDA could not take immediate, decisive action. [18]

The agency's only regulatory option was to attempt to revoke the New Animal Drug Application (NADA) of the twenty or so manufacturers of DES-laced cattle feed premixes. NADAs, a consequence of the Kefauver-Harris Amendment to the Food, Drug and Cosmetic Act (1962), placed on an animal drug's manufacturer the burden of demonstrating to the FDA's satisfaction its safety and efficacy. That obligation remained in force so long as the drug was used in the prescribed fashion, which required manufacturers to make their case successfully each time the FDA issued a new challenge. The revocation process began with an announcement in the *Federal Register*. The FDA set out the new charges against an animal drug and gave manufacturers a month to request a hearing to present evidence disputing the published contentions. If no hearing application was submitted within that time or if the FDA commissioner deemed not germane the evidence that the NADA holders outlined in their request for a hearing, the FDA would revoke the NADA. Use of the drug under conditions described in the NADA became illegal. When a hearing was granted, however, adjudication was much more deliberate, often taking several years. During this procedure, the drug could continue to be utilized. The matter went first before an administrative law judge, who conducted in effect a full-scale trial and issued a decision. These findings were then taken under advisement by the FDA commissioner, who again considered the evidence, consulted the appropriate specialists, and rendered in the form of a legal brief the ultimate verdict.[19]

The 1972 DES challenge never reached a hearing. Several NADA holders put forth what they considered relevant evidence and asked for a hearing, but Commissioner Charles E. Edwards, under heavy pressure from the public and from Congress, denied their application. On August 2 he prohibited further production of the cattle growth promoting substance. His move was less decisive than it first appeared, however. Maintaining that DES meat had never caused "a single known instance of human harm," the commissioner declared that "there is no justification for an abrupt disruption of the nation's meat supply" and, in what must have been a sop to premix manufacturers and cattlemen, permitted already produced DES premixes to be sold and used until January 1, 1973.[20]

The commissioner's ruling allowing DES use in cattle until the new year incensed those who had demanded its immediate withdrawal from the market, while his refusal to grant a hearing frustrated premix manufacturers. At the behest of L. H. Fountain (D-NC), who as chairman of the House Intergovernmental Relations Subcommittee had long opposed addition of DES to cattle feed, the Comptroller General of the United States examined the commissioner's phase-out tactics and proclaimed them illegal, but lacked authority to compel change. The Senate, which had had a bill outlawing DES pending when the commissioner issued his decision, took up the matter in earnest. Led by a coterie of liberals, including Edward M. Kennedy, Clifford Case, Abraham Ribicoff, Jacob Javits, and John Pastore of the heavily populated Northeast and Gaylord Nelson, Mike Mansfield, George McGovern, and William Proxmire of the beef-producing Midwest, the Senate quickly passed the anti-DES measure. The upper chamber's act proved merely symbolic as the House of Representatives never considered the matter. While avoiding public displays of dissatisfaction, the premix manufacturers had not sat idle. They had filed suit in district court, hoping to overturn the commissioner's ban because he had failed to provide a hearing.[21]

On January 24, 1974, the court ruled in favor of the plaintiffs, reinstated the NADAs, and permitted marketing and use of DES premixes. It chided the commissioner for attempting to decide the DES question "as a matter of paternalistic sagacity" and rejected his contention that "the court['s] sound course lies in deference to his expertise and judgment as the agent charged with responsibility for the public health." Noting that America was a nation of laws, the court determined it improper "to ignore procedural deficiencies on the basis of the public health hazard that has not been declared."[22]

In the wake of the court's decision to ignore the commissioner's protestations of expertise, the FDA began the time-consuming process of marshaling its facts in preparation for a hearing. Among its first acts was revocation of the mouse uterine bioassay as the official method of testing DES in beef. The commissioner claimed the test was insufficiently sensitive to serve as a regulatory tool and no doubt shocked premix manufacturers, animal scientists, cattlemen, and others by offering no new test, not even the erratic GLC, in its stead. Abolition of an official method prohibited supporters of DES's cattle growth promoting properties from employing an assay of inferior sensitivity to demonstrate the absence of the drug in meat and hence the safety of stilbestrol under present terms of use. At the heart of this move rested new USDA investigations then in progress at Fargo, North Dakota. USDA researchers used radioassays—radioactive tracer studies sensitive to parts per trillion or even more minuscule amounts of DES—to identify the presence of DES, its conjugates, or metabolites in livers of treated cattle a full seven days after the drug was last fed to these animals. This technique left them unable to determine the quantity of DES or its derivatives there, only that some incredibly small amount probably remained.[23]

These findings led not only to revocation of the official method but also to a lengthening of the mandatory withdrawal period to fourteen days. The FDA then retired temporarily from the scene to let Congress consider legislation banning DES in animal feed. Indeed, that course seemed the speediest and most prudent. The agency remained unable to justify the statutory conditions necessary to invoke either the Delaney cancer clause or the imminent hazard provision, and conventional regulatory proceedings might drag on for years. Congress received an additional incentive to move on the question on April 10, 1974, when the Canadian government effectively terminated sale of U.S. cattle and meat to Canada. It mandated that all imported beef and beef cattle must be certified by the exporting government as not having been treated with DES, a measure with which the U.S. government could not comply. The action of this largest U.S. beef trading partner was gutted four months later when voluntary compliance by individual feedlot operators replaced governmental oversight, but not before Canada's move had helped galvanize anti-DES sentiment in the Senate. On September 9, 1975, the U.S. Senate voted by a two-to-one margin to outlaw the drug, but the more conservative House never considered the Senate bill, as it remained bottled up in committee as the Ninety-fourth Congress adjourned.[24]

In early 1976 the FDA again initiated the long process required to revoke the diethylstilbestrol NADAs. Public pressure on the agency had intensified in the wake of congressional inaction. But the genesis of FDA's latest thrust came at least in part as a consequence of new studies suggesting that some DES sons and daughters suffered noncancerous lesions in or deformities of their urogenital tracts; DES seemed a transplacental mutagen. Far more common than incidents of cancer or precancerous conditions in DES daughters, which topped out at about two hundred cases in two million, these deformities and lesions were somatic, not carried to future progeny. The overwhelming majority were apparently harmless, mere asymptomatic clinical curiosities, although a few resulted in such disastrous conditions as infertility.[25] These studies did not dissuade premix manufacturers, however, and several asked for a hearing. They received backing in their opposition to a ban from the National Cattlemen's Association, the beef producers' trade organization, and the prestigious American Society of Animal Science, the nation's largest association of animal research scientists. The Council for Agricultural Science and Technology (CAST), a consortium of eighteen agriculture-related scientific societies—including the American Society of Animal Science—also threw its support behind continued DES use. In fact, CAST had been established in 1972 to respond to just this sort of challenge. It was founded by agricultural scientists who were upset by what they recognized as the uninformed nature of the consumer and environmental movements and were frustrated by the relative neglect of agricultural science questions in the National Academy of Sciences. CAST sought to serve as the "voice of agricultural science" and to "make agricultural scientists more effective sources of public information," pressing their case in a new, seemingly all too public arena.[26]

This time the FDA complied with requests for a hearing. In these sessions, the agency took dead aim at CAST. The FDA's attack undermined further the appearance of scientific exactitude that both CAST and the FDA had so desperately wanted to present and reaffirmed that the notion of a pristine, homogeneous scientific community perpetually resting above the fray was an idea whose time had passed. The focus was a report issued by a panel of scientists in CAST's name claiming that potential DES residues in meat were no more harmful to humans than naturally occurring estrogens in wheat germ and green vegetables. "Toxicity is a matter of dosage," maintained the CAST contingent. "All substances in large enough doses are toxic; in small enough dosage they show no significant activity." The FDA called the report a "conclusion in search of citations," one marked by "phantom references." Alluding to the fact that CAST received a significant portion of its funding from industry, the bureau reasoned that the industrial connection tainted the report, prejudicing its scientists and discrediting their analyses. What resulted from this unholy alliance was "a purportedly scientific document which in fact is a collection of conclusions by a number of cattle industry-oriented people followed by a list of disconnected references and appendices." [27]

That the FDA would espouse "a holier than thou" position on this ethical issue certainly was startling. The agency long had been attacked for having been staffed with former agricultural college or drug company scientists as well as for establishing advisory committees composed of similarly credentialed researchers. Those who launched these verbal onslaughts linked past employment with current outlook, believed that those pasts had biased the investigators in favor of manufacturers' objectives, and demanded unsullied experts without conflicts of interest. These critics ignored the obvious: that virtually all veterinary medical and toxicological experts worked at some time for pharmaceutical concerns, land-grant universities, or the USDA. The agency had characteristically deflected criticism from and defended its staff and advisors by referring to their professionalism and sense of public service, arguing that "their own high scientific integrity and concern for the promotion and preservation of health" would guarantee impartiality; they were scientists first and foremost. [28]

As the DES hearing progressed, the FDA suffered an assault from yet another quarter. A letter written by two researchers appeared in the *Journal of the American Medical Association* ridiculing the FDA and challenging its scientists' professional regulatory judgments. On this occasion the critique complained of the agency's seeming willingness to kowtow to public sentiment. Under the title "Money Causes Cancer: Ban It," the two investigators reported that they had induced cancer in thirty-five rats by inserting sterilized dimes in the animals' peritoneal cavities. While acknowledging that foreign body tumor genesis was a fact long established, the two scientists contended that they had performed the experiment to dramatize their case against the FDA. Specifically, they denounced the agency for its "inane pronouncements" about cancer-producing substances in food and the environment, calling those statements "this nonsense" and questioned the practicality, effectiveness, and scientific validity of the Delaney clause.[29]

Response to the letter was swift and predictable. The *Wall Street Journal* championed it as the first step in getting Americans "to talk more reasonably" about public health issues. The Federation of American Scientists, a group that dealt explicitly with problems of science and society, was less charitable. It chose to ignore the regulatory issues raised by the two scientific correspondents, focusing instead on what it recognized as the trivialization and politicization of science. The authors' "underlying mistake" was to use science "to make a political point," noted the politically active organization, which "undermined and discredited the legitimate experiments of their colleagues." The rats/dimes experiments proved only "the importance of insulating science done as science from politics masquerading as science."[30]

The DES hearing concluded in this atmosphere. After it had accorded the NADA holders their say, the FDA submitted to the administrative law judge in March 1978 a legal brief that was anything but brief, running more than 67,000 pages. Requirements that the FDA consider questions of both health and the quality of life contributed to the document's enormous size. The consequence of congressional statutes and judicial rulings, institutionalization of these quality of life issues were some of the tangible products of consumerism. The safety of a drug had been the agency's original and sole regulatory decision-making charge. By the late 1950s, however, the agenda had been expanded to include efficacy—it was not codified as law until 1962—while a decade or so later environmental impact was added. These initiatives were followed in the 1970s by requirements for cost/benefit analyses and economic impact statements. And the relative merits of these often competing issues were never clear. In the case of DES, for instance, the FDA had to recognize that the drug saved roughly five hundred pounds of feed per animal, which perhaps depressed feed prices but also released land that otherwise might have gone to raising feed and kept beef prices down. Similarly, the agency had to consider that keeping each animal for a shorter duration on an efficient feed/premix combination decreased the amount of excrement per animal, which seemingly reduced the waste disposal problem but might actually have exacerbated it; the waste of DES-fed cattle contained a potent hormone that may have required special handling. It also had to weigh the fact that DES beef was lower in fat and higher in protein than beef produced without the drug and, as a result, was of potentially greater benefit. The risks of heart disease and colon cancer were presumably lessened and the nutritional content of the meat was increased, and the agency needed to determine how these factors compared with a possible heightened cancer chance.

These were but a few examples of the types of questions that federal law required the FDA to contemplate, the answers to which neither the FDA nor anyone else knew. The situation was compounded by the single truly medico-scientific issue, namely: was DES in doses as minute as those sometimes found in beef dangerous as a carcinogen or mutagen? Certainly there was no evidence to sustain such a conclusion. Alternately, there was no evidence establishing what constituted a safe dose of DES in humans—could one molecule be the threshold amount?—or demonstrating the effects of minute quantities of the drug ingested daily over a lifetime.

The administrative law judge pondered these and other issues and on September 21, 1978 ruled for the FDA. He maintained that the NADA holders had not provided information sufficient to prove use of DES safe, a task made exceedingly difficult by the absence of an official method; to demonstrate that massive environmental or economic harm would occur as a consequence of its banning; or to document that economic and environmental benefits clearly outweighed possible health risks. The commissioner took the decision under advisement, sifted through the evidence, and on July 29, 1979, prohibited further use of stilbestrol in animal feed.[31]

The commissioner's order effectively ended the DES imbroglio. Its implementation was postponed several months as the commissioner, "sensitive to the possibility that an immediate ban of DES would result in economic disruption in the cattle production industry," sought to create "a reasonable period for transition from the use of DES to alternative growth promotants" and allowed cattlemen to exhaust existing stocks of the drug. A few feedlot operators continued after this grace period to add DES to cattle feed in violation of the law, and later research showed that early studies greatly exaggerated the drug's mutagenic properties, but the matter was closed. Cattlemen shifted to other estrogenic substances, such as estradiol, to boost cattle growth.[32]

With the stroke of a pen, the commissioner had answered the legal question of DES as a beef cattle booster. The nature of the complaints exposed during the DES debates as well as the act of complaining itself were not as easily resolved. Shearing decision making of the mask of objectivity had left individuals to their own devices. Each had assessed the evidence he or she had accumulated and evaluated it according to her or his individual standards and perspectives. Those reaching similar, personalized positions had joined together to attempt to influence or overwhelm the FDA, each member of which engaged in a process not unlike that undergone by many individuals outside government. But this phenomenon—rejection of scientific experts as disinterested arbiters—was not restricted to DES in meat or even the FDA. Rather, this rebellion against professionals, professionalism, and experts of every stripe was a characteristic American thread from the late 1950s. Indeed, it was a manifestation of the individuation of American society, an individuation that was reflected in virtually every sphere of activity. The then established consensual order began to show cracks in the 1950s as litterateurs and others launched an attack on authority. Volumes such as *The Organization Man* by William H. Whyte, Jr., David Reisman's *The Lonely Crowd,* and Sloan Wilson's *The Man in the Gray Flannel Suit* railed against suburbanization, conformity, corporatism, and scientism and pleaded with readers to assert their individuality. Glorification of the disaffected—manifested most clearly by the popularity of James Dean in *Rebel Without a Cause* and in his life—was followed by fears of rampant juvenile delinquency, allegedly furthered by individually held transistor radios playing rock and roll. Beatniks, the ultimate nonconformists, gave way to a bald rejection of established conventions and institutions in the 1960s. By the end of that decade, discourses of victimization and entitlement had appeared and would reach full flower in the 1970s and beyond. These discourses added immediacy and bitterness to the mix as individuals typically and repeatedly claimed that others injured, mistreated, or hampered them in their quest for self-fulfillment and that they had the right to achieve whatever end they demanded.[33]

It was the spurning of the pretense of consensus and replacement—legitimation, really—of individual perspective that produced newer forms of analysis—new "problems"—and, more important, the inability of those debating to agree on much of anything. How one conceives of a situation, how one approaches it and interprets it in a world in which expertise has been replaced by individual perspective, individual autonomy, and individual self-interest was the fundamental cultural predicament; it remains the defining event of late twentieth-century America. Means to "resolve" these new "problems" also stemmed directly from this individuation. Individuation of evaluation and rejection of the idea of consensual elements often led to formation of coalitions, which usually coalesced around some perceived fear or threat. Each issue-specific coalition was to come to bear on a single perceived problem, situation, or condition and to be sustained with its accepted authority only so long as that issue remained viable. Individuals retained their ability to withdraw at any time.

That these interest groups of self-interested individuals are singularly interested made negotiations toward a consensual resolution among interested groups difficult, if not impossible, on any issue. Their raison d'être was their position on the issue; to compromise was to undercut the group's reason for being. These singularly interested interest groups often identify themselves as "true" forms of identity, moreover, and then set out to try to give their group some kind of reality or grounding in reality, usually by seeking those social forces they claim created it. In scholarly circles, for example, ethnicity, gender, and class have become the self-defining attributes. In the area of governmental regulation, the ascendancy of individual prerogative generally received the designation "consumerism," a tacit acknowledgment both that those who viewed themselves primarily as purchasers and users of products constituted an interest group that manufacturers needed to satisfy and that this perspective was valid. In fact, consumerism emerged as such a potent concept in the 1970s that ranking federal regulatory agency officials sometimes were forced to spend more than one hundred working days each year testifying before congressional oversight committees.

Individuation was as fully pronounced within the sciences. Beneath surface protestations to the contrary, there emerged among scientists during the second half of the twentieth century a profound lack of agreement about what constituted science, although not about science's supremacy and potential utility. These men and women also questioned who were the experts—who were the "real" scientists—which served as a potent challenge to the idea of scientific specialties or disciplines. The percentage of scientists in traditional disciplinary or specialty organizations declined precipitously, while the number of splinter and otherwise fringe scientific interest groups increased explosively. Most important, peer review, the examination of a scientist's work by others before it achieved any sort of credibility or was fit for public scrutiny, seems to have broken down. Scientists regularly sought funding through direct congressional appropriation, not peer review–driven mechanisms. Heated public disputes among scientists, long a tacit violation of professional ethics, became de rigueur. Scientific journals, the essence of peer review, now published research papers long after the results had appeared in newspapers. It seemed as if the *New York Times* was on the verge of becoming the professional journal of choice.

Afterword

It would appear only reasonable, in a book that began with a series of methodological arguments for serious students of cultural history, to assess what the yield of the methods and approaches championed herein has been. And, although the number of possible perspectives is not infinite, nevertheless we would require far more than the limited space available here to suggest some of these points of view. Let us focus on one large theme: the changing meaning of individualism in American history. It may astonish, at first, that such an insight could arise from the study of technical knowledge—from, that is, science, technology, and medicine. But it is not so surprising once we realize that, of course, technical knowledge, like any other human discourse or action, is the product of what people say and do. Clearly technical knowledge could reveal much about the attitudes of a democratic culture toward one of its most cherished issues: the importance of the individual in society.

In the first three essays, we find an early to middle nineteenth-century American culture and society in transition from what most historians have called a republic to a democracy. In Spraul-Schmidt's essay, for example, we find that early nineteenth-century Americans believed in the virtues of individuals, who could improve themselves and make themselves legitimate members of civilized society through programs of self-improvement; any such individual could stand alone in a laissez-faire political economy and civilized society and culture. From the late 1830s on, however, as Spraul-Schmidt argues, the Ohio Mechanic's Institute changed its programs and purposes dramatically. No longer was it to create an independent civilized craftsman, small-scale entrepreneur, and head of household, which had been its original plan. Now it existed to train a class of mechanical operatives or, more precisely, various groups in the social hierarchy in various technical skills and ideas. Once the institute's trustees thought that society's basic social unit was the free, educable, rational individual who, once properly prepared, could stand alone as an independent entrepreneur and head of household in an atomistic, laissez-faire society. By the mid-nineteenth-century, however, they thought of society as comprising many distinct groups, racial, religious, occupational, and the like; now it made sense to prepare those various groups that did not fit into society's white, Anglo-Saxon, Protestant middle class for full participation in that class, including adherence to its cultural and social norms of behavior. In the essays of Lindee and Marcus we see the same kind of shift, from individual to group, and occurring at roughly the same time, the late 1830s and early 1840s, with profound consequences for the pursuit of science and the practice of medicine.

The next four essays explore the multiple meanings and nuances of the dynamics of group identity for the individual and group conflict within the larger society. Strictly speaking, from the later nineteenth to the mid-twentieth century, there was no such thing as individualism, save as a code of ethics or of discipline (as in the so-called Puritan work ethic) intended to make an individual conform to preexisting group norms. This part of the book begins with Theriot's assessment of the strange affliction—called puerperal insanity—that infected new mothers and became a part of the interaction of the various interested parties (dare we say groups?) and was thus socially constructed, to use a currently fashionable verb, by particular groups and interests. That this affliction disappeared by the second decade of the twentieth century is indeed suggestive that it was the construction not so much of this social interest or that, but of a pervasive sense of the order of things for a particular age. Similarly, the reader can compare and contrast the essays of Spraul-Schmidt and Layton to see the rise and fall of the independent mechanic in nineteenth-century American culture. As the James Emersons of America fell victim to the new class of scientifically trained engineers, one could see the fading importance of such institutions as the Ohio Mechanic's Institute and, for that matter, the changing meaning of "mechanic," which, as Layton points out, paralleled the changing meaning of "yeoman farmer," signaling the decline of the independent entrepreneur in the shop and on the farm and the rise of the wage-earning operative, as disposable as yesterday's paper napkins at the dinner table. Or so it appeared to many contemporary Americans. Such was a slender basis indeed upon which to lead a life of cultural, social, political, and economic independence, as many Americans had imagined that independent mechanics and yeoman farmers had in American history. The essays by Miller and Cravens on unusual conceptions and definitions of racial analysis confirm the pervasiveness of racial and group analysis in rather unlikely quarters, namely the ideas of the great African-American intellectual William E. B. Du Bois and those of a variegated collection of scientists and politicians seemingly interested in the scientific and social problems of human nutrition. Yet racial analysis, and the assumption that group identity for the individual was inescapable, pervaded these two case studies.

The final group of three essays documents the shift in American culture and society from group to individual. Cravens again contributes an essay, this time on the problems of conceptualizing the intelligence quotient as a measure of group or individual performance, which confounded and befuddled an entire scientific profession. Fairbanks treats in splendid fashion the changing definition in the same era of the basic unit of society from community to individual in his study of rivalry between Dallas and Fort Worth over the placement of a major airport. Marcus returns to the discussion with a sophisticated essay on the evaporation of the once widely shared assumptions necessary for Americans to accept expertise as a unified and impartial authority, as distinct from the post-1950s attack on expertise and the rise of competing expertises, one for each interest group in any competition in social and public policy.

So it would appear that American culture and society have gone full circle, from basing society on the individual to the group to the individual. By this we do not mean that history repeats itself. Such a generalization is too broad to give us much insight, for each age in the past is unique and stands on its own. Rather, it would appear that the individual and the group have functioned as fundamental building blocks in American (and European, most likely) social imagination, in those blueprints which contemporaries use to describe and interpret the world around them. And thus we have bridged the gap between culture and technical knowledge and produced insights of considerable value and interest. We hope that students of cultural history who read this book will be inspired to follow in their own work the methodological ideas we have suggested here. We are confident that they too will produce exciting—and historically sound—insights into that most foreign of all foreign countries, the past.

Notes

Introduction

1. Stow S. Persons, ed., *Evolutionary Thought in America*. Special Program in American Civilization at Princeton (New Haven: Yale University Press, 1950); Persons, *American Minds: A History of Ideas* (New York: Holt, Rinehart, and Winston, 1938); Merle Curti, *The Growth of American Thought* (New York: Dodd and Mead, 1943); Curti, *Human Nature in American History: A History* (Madison: University of Wisconsin Press, 1980).

2. Thomas S. Kuhn, *The Structure of Scientific Revolutions*, 2d ed., enlarged (Chicago: University of Chicago Press, 1970); Alexander Koyre, *From a Closed World to an Infinite Universe* (Baltimore: Johns Hopkins University Press, 1957). A useful starting point for historiographic and bibliographic assistance in history of science, medicine, and technology is Paul T. Durbin, ed., *A Guide to the Culture of Science, Technology and Medicine* (New York: Free Press, 1980).

3. Henry Nash Smith, *Virgin Land: The American West as Symbol and Myth,* twentieth anniversary reissue (Cambridge: Harvard University Press, 1970), xi. Leo Marx, *The Machine in the Garden: Technology and the Pastoral Ideal in America* (New York: Oxford University Press, 1964); R. W. B. Lewis, *The American Adam* (New York: 1955); John William Ward, *Andrew Jackson: Symbol for an Age* (New York: Oxford University Press, 1955); Marvin Meyers, *The Jacksonian Persuasion: Politics and Belief* (Stanford: Stanford University Press, 1957).

4. Marx, *Machine in the Garden*, quotes at 354, 226; Marvin Fisher, *Workshops in the Wilderness: The European Response to American Industrialization, 1830–1860* (New York: Oxford University Press, 1967).

5. For a useful discussion of these attitudes, see Alan I Marcus and Howard P. Segal, *Technology in America: A Brief History* (San Diego: Harcourt, Brace, Jovanovich, 1989), 257–310.

6. See, for example, Gene Wise, " 'Paradigm Dramas' in American Studies: A Cultural and Institutional History of the Movement," *American Quarterly* 31 (Summer 1979): 292–337; Wise, *American Historical Explanations: A Strategy for Grounded Inquiry* (Chicago: Dorsey, 1973).

7. Bruce Kuklick, "Myth and Symbol in American Studies," *American Quarterly* 24 (October 1972): 435–50.

8. Howard P. Segal, "Leo Marx's 'Middle Landscape': A Critique, a Revision, and an Appreciation," *Reviews in American History* 5 (March 1977): 137–50.

9. A good general guide to the literature of the history of science, medicine, and technology in the United States is Sally G. Kohlstedt and Margaret W. Rossiter, eds., *Historical Writing on American Science: Perspectives and Prospects* (Baltimore: Johns Hopkins University Press, 1986). Examples of this perspective are David F. Noble, *America by Design: Science, Technology, and the Rise of Corporate Capitalism* (New York: Oxford University Press, 1977); Merritt Roe Smith, *Harpers Ferry Armory and the New Technology: The Challenge of Change* (Ithaca: Cornell University Press, 1978).

10. An examination of the September 1989 issue of the *Journal of American History* (vol. 76, no. 2) will provide accurate and revealing statements of this agenda.

11. On our own era, see, for example: Peter Clecak, *America's Quest for the Ideal Self: Dissent and Fulfillment in the '60's and '70's* (New York: Oxford University Press, 1983); Alan I Marcus, "The Wisdom of the Body Politic: The Changing Nature of Publicly Sponsored American Agricultural Research Since the 1830s," *Agricultural History* 62 (Spring 1988): 4–26; Marcus and Segal, *Technology in America,* 315–61; Hamilton Cravens, "History of the Social Sciences," *Osiris,* second series I (1985): 183–207; Cravens, "Recent Controversy in Human Development: A Historical View," *Human Development* 30 (December 1987): 325–35; William H. Whyte, Jr., *The Organization Man* (New York: Simon and Schuster, 1956). Anne Moody, *Coming of Age in Mississippi* (New York: Dell Books, Inc., 1968), is a wonderful text for the 1960s. Probably the text most representative of American culture in the 1970s is George M. Sheehan, *Running and Being: The Total Experience* (New York: Warner Books, 1978), and a good choice for the 1980s is George Gilder, *The Spirit of Enterprise* (New York: Simon and Schuster, 1984).

12. For a recent and widely noted example of this kind of thinking, see Thomas Bender, *New York Intellect: A History of Intellectual Life in New York City, from 1750 to the Beginnings of Our Own Time* (New York: Knopf, 1987).

13. Excellent examples of these classic works in American intellectual history are Curti, *The Growth of American Thought;* Persons, *American Minds;* Richard Hofstadter, *The American Political Tradition and the Men Who Made It* (New York: Knopf, 1948); Ralph Henry Gabriel, *The Course of American Democratic Thought* (New York: Ronald Press, 1940); Eric Goldman, *Rendezvous With Destiny* (New York: Knopf, 1953), Gordon S. Wood, *The Creation of the American Republic* (New York: Knopf, 1969); Linda K. Kerber, *Women of the Republic* (Chapel Hill: University of North Carolina Press, 1980); Carl N. Degler, *In Search of Human Nature* (New York: Knopf, 1991).

14. Stephen J. Gould, *The Mismeasure of Man* (New York: W. W. Norton and Company, 1981). Among the best examples of the latter social history of science school has been the work of Charles Rosenberg, which has been enormously influential in the profession; of all of Rosenberg's work perhaps nothing is more

widely cited and used than his *The Cholera Years: The United States in 1832, 1849, and 1866* (Chicago: University of Chicago Press, 1962).

15. Brooke Hindle, *The Pursuit of Science in Revolutionary America* (Chapel Hill: University of North Carolina Press, 1956); Daniel J. Boorstin, *The Lost World of Thomas Jefferson* (New York: Henry Holt and Co., 1948); George H. Daniels, *American Science in the Age of Jackson* (New York: Columbia University Press, 1967); Henry D. Shapiro, *Appalachia on Our Mind: The Southern Mountains and Mountaineers in the American Consciousness* (Chapel Hill: University of North Carolina Press, 1978).

I. The Ohio Mechanic's Institute

1. "An Act to Incorporate the Ohio Mechanic's Institute," *Laws of Ohio*, 1829. See Alice Felt Tyler, *Freedom's Ferment: Phases of American Social History to 1860* (Minneapolis: University of Minnesota Press, 1944; New York: Harper Torch books, 1962), 227–50 and 262 for encyclopedic coverage of the associational proliferation and the lyceum quote. Also see Carl Bode, *The American Lyceum: Town Meeting of the Mind* (New York: Oxford University Press, 1956). On gentility, see Stow Persons, *The Decline of American Gentility* (New York: Columbia University Press, 1973).

2. Henry D. Shapiro, *Appalachia on Our Mind: The Southern Mountains and Mountaineers in the American Consciousness, 1870–1920* (Chapel Hill: University of North Carolina Press, 1978) lays out the conception of "history as problem solving" in its preface. Shapiro's specific ideas on the early nineteenth century expressed in his 1976 University of Cincinnati Bicentennial Lecture and other places also have been a significant influence. For the definition of "science," see Bruce Sinclair, *Philadelphia's Philosopher Mechanics: A History of the Franklin Institute, 1824–1865* (Baltimore: Johns Hopkins University Press, 1974), 2.

3. Alan I Marcus, "In Sickness and in Health: The Marriage of the Municipal Corporation to the Public Interest and the Problem of Public Health, 1820–1870" (Ph.D. diss., University of Cincinnati, 1979) lays out in detail this conception of the early nineteenth century as an "age of the individual" in which "the individual was the basic unit of society." Marcus was extremely helpful in this essay's formulation. Also see his *Plague of Strangers: Social Groups and the Origins of City Services in Cincinnati, 1819–1870* (Columbus: Ohio State University Press, 1991). For the quotes, see Tyler, *Freedom's Ferment*, 277–78 and 303. Barbara J. Berg, *The Remembered Gate* (New York: Oxford University Press, 1978) notes that "female associations proliferated in the years after 1820" and gives examples. David J. Rothman, *The Discovery of the Asylum: Social Order and Disorder in the New Republic* (Boston: Houghton Mifflin, 1971) notes the creation of different types of asylums as a response to disorder in the era; see also Fred Somkin, *Unquiet Eagle: Memory and Desire in the Idea of American Freedom, 1815–1860* (Ithaca: Cornell University Press, 1967).

4. Zane L. Miller, "Scarcity, Abundance, and American Urban History," *Journal of Urban History* 4 (February 1978): 131–55 discusses cities in early America in a very useful manner. See pp. 137 and 139 for quotations. Of especial note is Alan I Marcus, "The Strange Career of Early Nineteenth Century City Government: The Case of Cincinnati," *Journal of Urban History* 7 (November 1980):

3–29; Richard C. Wade, *The Urban Frontier: Pioneer Life in Early Pittsburgh, Cincinnati, Lexington, Louisville, and St. Louis,* (Cambridge: Harvard University Press, 1959) is also important. See also Judith Spraul-Schmidt, "The Origins of Modern City Government: From Corporate Regulation to Municipal Corporation in New York, New Orleans and Cincinnati, 1785–1870," (Ph.D. diss., University of Cincinnati, 1990). Examples of states and cities seeking to increase their hinterlands and influence through transportation strategies include Robert G. Albion, *The Rise of New York Port, 1815–1860,* (New York: Charles Scribners' Sons, 1939); and Harry N. Scheiber, *Ohio Canal Era: A Case Study of Government and the Economy, 1820–1861,* (Athens: Ohio University Press, 1969).

5. Miller, "Scarcity"; Wade, *Urban Frontier;* Daniel Drake, *Natural and Statistical View or Picture of Cincinnati and the Miami Country* (Cincinnati, n.p., 1815); B. Drake and E. D. Mansfield, *Cincinnati in 1826* (Cincinnati: n.p., 1826). Mechanics' institutes were established in New York in 1823, Philadelphia in 1824, Baltimore in 1825; and Boston in 1827. Institutions of the same name were formed in Glasgow in 1823 and London in 1824, but these differed from the American counterparts in demonstrating a certain class consciousness. See Sinclair, *Philadelphia's Philosopher Mechanics,* 1–19. For the Cincinnati quote, see *Cincinnati Daily Gazette,* October 27, 1828, p. 2.

6. *Cincinnati Daily Gazette,* October 27, 1828, p. 2; Wade, *Urban Frontier;* and Zane L. Miller, "Cincinnati: A Bicentennial Reassessment," *Cincinnati Historical Society Bulletin* 34 (1977): 34–53.

7. *Cincinnati Daily Gazette,* November 19, 1829; and John D. Craig, "An Address Delivered at a Meeting of the Citizens of Cincinnati Convened for the Purpose of Forming a Mechanic's Institute" (Cincinnati, 1828), 1, 6, 9, 16. For a brief biography of John D. Craig, see Sinclair, *Philadelphia's Philosopher Mechanics,* 14, 135.

8. Craig, "Address"; Craig's musings, especially 5 and 9, evoke Sinclair's summary of Edward Everett, "On the Importance of Scientific Knowledge to Practical Men, and on the Encouragement to its Pursuits." This and other information suggests that the early years of the Ohio Mechanic's Institute and the Franklin Institute were more alike than Sinclair has understood. Also John William Ward, *Andrew Jackson: Symbol for an Age* (New York: Oxford University Press, 1955) and Marvin Meyers, *The Jacksonian Persuasion: Politic and Belief* (Stanford: Stanford University Press, 1957).

9. *Liberty Hall,* November 9, 1828; "Constitution of the Ohio Mechanic's Institute, 1828," Ohio Mechanic's Institute Papers, Special Collections, (hereinafter referred to as the OMI Papers) housed at the University of Cincinnati Libraries; CHS Pamphlet Collection, housed at the Cincinnati Historical Society.

10. Miller, "Scarcity," 153; For early nineteenth-century corporations, see Nelson M. Blake, *Water for the Cities* (Syracuse: Syracuse University Press, 1956). On the broad sense of the term "mechanic" in this period, see Monte A. Calvert, *The Mechanical Engineer in America, 1830–1910* (Baltimore: Johns Hopkins University Press, 1967), 29, 44. Calvert maintains that as late as 1850, the term "mechanic" did not suggest mechanical engineer.

11. *Cincinnati Daily Gazette,* March 16, 1829; Sinclair, *Philadelphia's Philosopher Mechanics,* also notes that the Franklin Institute membership in the 1820s and 1830s included type founders and printers, architects, chemists, saddlers, tailors, coppersmiths, glaziers, engine makers, bakers, painters, gunsmiths, doctors, lawyers, cordwainers, joiners, smiths, dyers, and carpenters; see p. 41. Indeed, the Franklin

Institute was founded by a commercial man turned manufacturer of fire engines, and a professor of chemistry.

12. *Cincinnati Daily Gazette,* February 21, 1829, p. 3; April 16, 1829. George W. Kendall, *A Sketch of the History of the Ohio Mechanics Institute and Statement of Its Present Condition* (Cincinnati: n.p., 1853) and Sinclair, *Philadelphia's Philosopher Mechanics,* 14. For a brief biography of John Locke, see *Dictionary of American Biography,* vol. 9.

13. *Cincinnati Daily Gazette,* December 1829.

14. Craig, "Address," 11; *Cincinnati Daily Gazette,* January 14, 1831, p. 2; see also Sinclair, *Philadelphia's Philosopher Mechanics,* as well as George Daniels, *American Science in the Age of Jackson* (New York: Columbia University Press, 1968).

15. "The objects of both [the Institute and the Lyceum] are the same—the general diffusion of knowledge, and the development of the mental and moral powers," *Cincinnati Daily Gazette,* December 7, 1832. See also May 4, 1831; February 19 and May 5, 1832; January 17 and February 7, 1835; January 9, 1836; June 6, 1840; and May 29 and July 23, 1841.

16. Henry D. Shapiro and Zane L. Miller, eds., *Physician to the West: Selected Writings of Daniel Drake on Science and Society* (Lexington: University Press of Kentucky, 1970). For the quote, see *Cincinnati Daily Gazette,* March 16, 1829; for the ladies, see July 9, 1840, p. 2 and May 29, 1841, p. 2.

17. *Cincinnati Daily Gazette,* March 23, 1829, statement issued by D. T. Disney, March 21, 1829, as secretary of the Mechanic's Institute.

18. "Rules for the Reading Room," in Minutes of the Board of Directors of the Ohio Mechanic's Institute, OMI Papers. The Philosophical Society of Cincinnati records are also in the OMI Papers. See also *Cincinnati Daily Gazette,* December 25, 1830, and Constitution of the Ohio Mechanic's Institute.

19. Constitution of the Philosophical Society of Cincinnati, OMI Papers. See also Henry D. Shapiro, "Daniel Drake's Sensorium Commune and the Organization of the Second American Enlightenment," *Cincinnati Historical Society Bulletin* 27 (1969): 43–52.

20. *Cincinnati Daily Gazette,* February 19, 1831; and "By-Laws For Regulating the Library, 1831," OMI Papers.

21. "By-Laws"; Kendall, "A Sketch of the OMI"; *Cincinnati Daily Gazette,* February 2, 1833, p. 3; November 29, 1832, p. 3; and December 25, 1830, p. 3.

22. *Cincinnati Daily Gazette,* March 11, 12, 14, 1833.

23. Report of the First Annual Fair, OMI Papers. See also *Cincinnati Daily Gazette,* February 15, 24, 26, 28 and March 7, 13, 1838, for ads and commentary on the ball; see too Kendall, "A Sketch of the OMI."

24. New York, Baltimore, and Philadelphia had all offered exhibitions. See Sinclair, especially 82–108, on the Franklin Institute exhibitions, 1824–38; see also Samuel Rezneck, "The Rise and Early Development of Industrial Consciousness in the United States, 1760–1830," *Journal of Economic and Business History* 4 (August 1932): 784–811 for a discussion of early exhibitions in a framework different from this one. *Cincinnati Daily Gazette,* 1829: August 16, p. 1; September 12, 2; October 28; November 27, 2, and December 29, 2.

25. "Report of the First Annual Fair," OMI Papers; Circular of the First Annual Fair, CHS Pamphlet Collection; and "Judges' Reports" (some printed and some manuscripts), OMI Papers; and *Cincinnati Daily Gazette,* January 29, May 9, 10, 12, and June 1, 1838.

26. "Report of the First Annual Fair": 6, 27, 46, 48, 50.

27. *Proceedings of the Second Annual Fair of the Ohio Mechanic's Institute* (Cincinnati, 1839), 27, 29; "Report of the Committee of Arrangements," 3–4, 6–24; *Cincinnati Daily Gazette,* June 8 and 21, 1839. See also Reports of the Third (1840), Fourth (1841), Fifth (1842), Sixth (1843), Seventh (1844), and Eighth (1845) Annual Fairs, OMI Papers.

28. For the quote, *Cincinnati Daily Gazette,* July 21, 1841, p. 3; "Minutes of the Third Fair Committee, January 18, 1840," OMI Papers; "Report of the Third Annual Fair," 1–4, 34–39; *Cincinnati Daily Gazette,* May 23, June 4, 23, 27, July 9, 10, 11, 13, 14, 18, 1840; *Spirit of the Times,* April 25, 1840; Minutes of the Committee of Arrangements of the Fourth Annual Fair, OMI Papers, especially March 27, April 10 and 15, 1841; *Cincinnati Enquirer,* April 10, 1841; "Report of the Fourth Annual Fair"; *Cincinnati Daily Gazette,* July 5, 1841; "Circular of the Fifth Annual Fair," OMI Papers, Box 83. One of the judges' reports was written on the reverse of a circular, and hence it was preserved. The file "Judges' Reports, Fifth Annual Fair," OMI Papers, Box 83 also holds manuscript judges' reports for 1843, 1844, and 1845. For an articulation of Cincinnati's aspirations to metropolitan status, see Charles Cist, *Cincinnati in 1841* (Cincinnati: n.p., 1841) and *Cincinnati in 1851* (Cincinnati: n.p., 1851); and Miller, "Cincinnati: A Bicentennial Reassessment."

29. W. T. Howe, "Address at the eighth annual fair, "Report of the Eighth Annual Fair," p. 5, OMI Papers; "Report of the First Annual Fair"; Miller, "Cincinnati: A Bicentennial Reassessment"; Michael Chevalier, *Society, Manners, and Politics in the United States* (1839; reprint, Garden City, N.Y.: Doubleday, 1961), 186, 198; and Craig, "Address."

30. The program continued to change over time. The institute sponsored a different series of annual exhibitions between 1850 and 1860, which continued to display a variety of products but categorized them with machinery the most important. In 1856 the institute organized a "School of Design" in which classes in design training were arranged into a specific program with gradations leading to the "more important branches" of "Mechanical" and "Architectural drafting." Minutes, Annual Reports, OMI Papers.

31. As industry assumed a new importance in the national economy, successful manufacturers began to claim a position of respect commensurate with their wealth, alongside merchants. In this context, the Mechanics Institute was oriented more directly toward the encouragement and advancement of industrial production. This change in thinking affected Cincinnati as it did cities in general, reinforcing the conviction that America's abundant resources could be expanded artificially, through human work, without limitation. But the habit of thinking in terms of groups had ramifications for cities themselves, as residence in any of them came to be itself a way of defining aggregates of individuals. The assumption of municipal responsibility for a whole range of public services showed the force of a new and fundamentally different definition of urban community. Cincinnati, which had risen rapidly from a turn of the century settlement to a bustling river town and civic center of the trans-Appalachian region, saw its chance to be not only the premier city of Ohio (as attested to in the 1828 naming of the Institute), or even the principal city of the West, as it called itself in the annual fairs, but a competitor for national metropolitan status.

32. Compilation of the officers, OMI Collection, University of Cincinnati.
33. *Cincinnati Daily Gazette,* July 21, 1841. For construction campaign, see OMI Papers, Box 11, folder 2, and bound vol. 1.
34. OMI Papers, vol. 1, pp. 338–39, subscriptions.
35. *Oration by E. L. Magoon delivered July 4th 1848 at Laying of the Cornerstone of the Ohio Mechanics Institute* (Cincinnati: n.p., 1848).

2. The American Career of Jane Marcet's *Conversations on Chemistry*

1. H. J. Mozans [John Zahm], *Woman in Science* (New York: Appleton, 1913), 372–73; Eva V. Armstrong, "Jane Marcet and Her Conversations on Chemistry," *Journal of Chemical Education* 15 (1938): 53–57. L. Pearce Williams gives Marcet's work credit for leading Faraday to connect his fascination with electrical phenomena with forces of "fundamental importance in the universe" chemical reactions; see Williams, *Michael Faraday: A Biography* (New York: Simon and Schuster, 1965), 18–20.
2. John K. Crellin, "Mrs. Marcet's *Conversations on Chemistry,*" *Journal of Chemical Education* 56 (1979): 459–60.
3. Alexander Marcet was a physician and, later, chemistry professor at Guys Hospital, London. When his wife inherited a substantial fortune upon the death of her father in 1817, Marcet was able to give up medicine and devote himself to chemistry. He was the author of several scientific papers, and his work on the specific heats of gases was cited in other textbooks and in his wife's book. Marcet's social circle is briefly explored in Armstrong, "Jane Marcet." See also Auguste de la Rive's obituary notice for Jane Marcet, "Madame Marcet," *Bibliothéque Revue Suisse et Etrangère* n.s., 4 (1859): 445–68 (transcription and translation in the Edgar Fahs Smith Collection, University of Pennsylvania, Philadelphia). Alexander Marcet is mentioned in J. R. Partington, *A History of Chemistry* (London: Macmillan, 1972); see also John Read, *Humour and Humanism in Chemistry* (London: Bell, 1947), 177–91, for some biographical details of Marcet's life.
4. Bryan's works included *A Compendious System of Astronomy* (London: n.p., 1797); *A Comprehensive Astronomical and Geographical Class Book* (London: n.p., 1815); and *Lectures on Natural Philosophy* (London:, n.p., 1806).
5. Jane Marcet, *Conversations on Chemistry* (Hartford: n.p., 1839), 150 (Caroline's father).
6. 1806 Philadelphia edition, 81 (on the well-informed); 1822 Hartford edition, 116, and 1839 Hartford edition, 125 (on colliers); and 1829 Hartford edition, 162 (on oxydate vs. rust).
7. Crellin, "Mrs. Marcet's *Conversations,*" cites a December 1803 letter from the London physician John Yelloly to Alexander Marcet, in which Yelloly seems to imply that Alexander is responsible for the quality of Jane's (as yet unpublished) manuscript.
8. 1839 Hartford edition, 14.
9. A.-L. Lavoisier, *Principles of Chemistry in a New Systematic Order,* trans. from the 2nd French ed. (1793) by Robert Kerr (Edinburgh: n.p., 1796). David Knight's comparison of Marcet's chemistry and that of Samuel Parkes, a contemporaneous writer who wrote for young men, emphasizes Marcet's theoretical assumptions; see Knight, "Accomplishment of Dogma: Chemistry in the Introductory Works of Jane Marcet and Samuel Parkes," *Ambix* 33 (1988): 94–98.

10. Thomas Thomson, *A System of Chemistry of Inorganic Bodies* (Edinburgh: n.p., 1831), 3–31; W. T. Brande, *A Manual of Chemistry* (London: n.p., 1841), 234–38; and Andrew Ure, *A Dictionary of Chemistry* (London: n.p., 1831), 443–45. I am indebted to J. R. Clarke for these useful comparisons, explored in his unpublished paper "Jane Marcet and Her *Conversations on Chemistry*" (J. R. Clarke, 15 Exeter Grove, Belmont, Victoria 3216, Australia).

11. See Lewis C. Beck, *A Manual of Chemistry* (Albany, N.Y.: n.p., 1831); John Lee Comstock, *Elements of Chemistry* (New York: n.p., 1839); and John Johnston, *A Manual of Chemistry* (Philadelphia: n.p., 1848, 1861).

12. Cf. Marcet, 1839 Hartford edition, conversations 2 and 3, with Edward Turner, *Elements of Chemistry* (n.p., 1842), 9–50.

13. Clarke makes this comparison in "Marcet and Her *Conversations*."

14. The book was often printed more than once in a single year by competing publishers, e.g., by Increase Cooke of New Haven and James Humphreys of Philadelphia in 1809. Two or more runs of Marcet's book or the imitative Thomas Jones, *New Conversations on Chemistry* were also produced by various American publishers in 1818, 1824, 1831, 1836, 1839, and 1844. Cornell's efficient Olin Library staff assisted me in tracking down sixteen of these twenty-three American editions, held in various libraries in the United States, and four editions of the version by Thomas Jones. I owe thanks also to the libraries that made copies available: University of Pennsylvania, University of Michigan, Princeton University, New York State Library at Albany, and University of Minnesota.

15. The book's popularity in Britain has been dismissed by some historians as a by-product of Humphry Davy's charisma; see Judith Brody, "The Pen Is Mightier Than the Test Tube," *New Scientist* 14 (February 1985): 58. See also Knight, "Accomplishment of Dogma," 97.

16. Jean Jacques has noted the anonymous publication in Paris in 1826 of *Entretiens sur la chimie après les méthodes of MM. Thenard et Davy*, virtually a direct translation of Marcet's text. Mrs. B. became Mme de Beaumont, Emily was transformed to Gustave, but Caroline remained Caroline. The same year A. Payen produced a version under the title *La chimie enseigneé en vingt-six leçons*. Though he restyled portions of the text, he lifted the order of the conversations and many discussions directly from Marcet's text. Jean Jacques, "Une chimiste qui avait de la conversation: Jane Marcet (1796–1858)," *Nouveau Journal de Chimie* 10 (1986): 209–11.

17. A modern facsimile of the German edition was published in 1984, with an afterword by the historian of chemistry Otto Paul Kratz (*Unterhaltungen ueber die Chemie*, trans. F. F. Runge [Wein-heim: Verlag Chemie, 1984]). Kratz concluded that the book failed in Germany because it discussed technologies, such as steam engines, unfamiliar to German readers. Karl Hufbauer has suggested that it may not have been successful because young German women had limited access to chemical education: see Hufbauer's review, *CHOC News* 2, no. 1 (Spring 1984): 7–8, on 8.

18. It is difficult to determine exactly how many copies of Marcet's work sold in America. Sarah J. Hale provided the figure of 160,000, cited by several other historians, in Sarah Josepha Buell Hale, *Woman's Record, or, Sketches of Distinguished Women from the Creation to A.D. 1868, Arranged in Four Eras* (New York:

n.p., 1874), 732. But Hale refers to 160 impressions of the book and assumes a print run of 1,000 copies per impression. I have managed to find records of only thirty-two impressions (counting Jones's version). Hale may have known of more, or perhaps print runs were much larger.

19. Thomas Jones included Marcet's name in the frontispiece of all his editions, but the text was listed as his work in catalogues of texts used in chemical instruction in the women's academies. See Thomas Woody, *A History of Women's Education in the United States* (Lancaster, Penn.: Science Press, 1929), 553. Woody also lists John Lauris Blake, the Episcopalian minister who provided questions for numerous editions, as the author of a chemistry text that must have been Marcet's work: Blake's questions appeared in every American edition after 1828. (Woody again lists Blake as the author of a natural philosophy text that must have been Marcet's later *Conversations on Natural Philosophy*, to which he added similar questions.) Blake also appears as the author of *Conversations on Chemistry* in the *Dictionary of American Biography* (1936): vol. 11, 343, and the *National Cyclopedia of American Biography* 21 (1931): 172. This attribution of male authorship must have been the work of persons who had not read the book, since every edition carried Marcet's self-deprecating preface, which was clearly written by a woman, and was almost unchanged for forty-four years.

20. Protection of copyright for American authors was established in 1790, but foreign authors were granted no such protection until 1891. The publication of Marcet's book in America was also influenced by the chaotic and competitive nature of early nineteenth-century book publishing. The cutthroat conditions of the era forced many publishers to specialize in fields where competition was not so general and returns more stable, such as science: Henry Walcott Boynton, *Annals of American Bookselling 1638 to 1850* (New York: John Wiley, 1932), 144. See also Warren S. Tryon and William Charvat, eds., *The Cost Books of Ticknor and Fields* (New York: Bibliographical Society of America, 1949), a reprint of the publishing records of a major Boston publisher from 1832 to 1858.

21. For a discussion of American envy of European chemistry, see Robert V. Bruce, *The Launching of Modern American Science, 1846–1876* (New York: Knopf, 1987), 14–28.

22. *National Cyclopedia of American Biography* 21 (1931): 172.

23. On Comstock, see *Dictionary of American Biography* (1872), 211; see also the extensive list of Comstock's publications in John F. Ohles, *Biographical Dictionary of American Educators,* vol. 1 (Westport, Conn.: Greenwood, 1978), 295.

24. John Lee Comstock, *A Grammar of Chemistry* (Hartford, Conn.: S. G. Goodrich, 1822), title page.

25. Keating was founder of the Franklin Institute and a chemistry professor at the University of Pennsylvania. Cooper was the son-in-law of Joseph Priestley, a professor of chemistry and mineralogy at the University of Pennsylvania and, after 1821, a chaired professor of chemistry at South Carolina College.

26. 1829 Hartford edition, 225.

27. 1825 Hartford edition, 13, 172.

28. 1839 Hartford edition, 281 (water), 234 (combustion).

29. 1822 Hartford edition, 79.

30. This introduction, essentially unchanged, is printed in every impression of Comstock's version before the table of contents.

31. Thomas Cooper, 1818 Philadelphia edition, preface; see also William H. Keating, 1824 Philadelphia edition.

32. See Wyndham Miles, "William H. Keating and the Beginning of Chemical Laboratory Instruction in America," *Library Chronicle* 29 (1952–53): 1–34. See also the entry on Keating in *Dictionary of American Biography* (1872), 502.

33. Wyndham D. Miles, "Public Lectures on Chemistry in the United States," *Ambix* 15 (1968): 129–53.

34. Thomas Jones, *New Conversations on Chemistry* (Philadelphia: n.p., 1832), preface.

35. Woody, *History of Women's Education,* vol. I, 553; Deborah Jean Warner, "Science Education for Women in Ante-Bellum America," *Isis* 69 (1979): 58–67; Linda K. Kerber, *Women of the Republic: Intellect and Ideology in Revolutionary America* (Chapel Hill: University of North Carolina Press, 1980); Anne Firor Scott, "The Ever-Widening Circle: The Diffusion of Feminist Values from the Troy Female Seminary, 1822–1872," *History of Education Quarterly,* Spring 1979, 3–25; and Patricia Cline Cohen, *A Calculating People: The Spread of Numeracy in Early America* (Chicago: University of Chicago Press, 1982), 134–49.

36. See the discussion in Paul J. Fay, "The History of Chemistry Teaching in American High Schools," *Journal of Chemical Education* 8 (1931): 1533–62, 1539–40. For a valuable review of the state of chemical instruction in American colleges see Bruce V. Lewenstein, "To Improve Our Knowledge in Nature and Arts: A History of Chemical Education in the United States," *Journal of Chemical Education* 66 (1989): 37–44.

37. Warner, "Science Education for Women," 59, 60 (quotation from Abbott Collegiate Institute, Catalogue [1854]).

38. Warner, "Science Education for Women," 62.

39. Almira Hart Lincoln Phelps, *Familiar Lectures on Chemistry* (New York: n.p., 1838). Besides works cited in notes 11, 43, and 44, I have considered the following texts: William Henry, *An Epitome of Chemistry* (Boston: n.p., 1810); John White Webster, *Manual of Chemistry* (Boston: n.p., 1826); Edward Turner, *Elements of Chemistry* (New York: n.p., 1828); James Renwick, *First Principles of Chemistry* (New York: n.p., 1840); Benjamin Silliman, *First Principles of Chemistry* (Philadelphia/Boston: n.p., 1847); Edward Youmans, *A Class Book of Chemistry* (New York: n.p., 1851); and Youmans, *The Handbook of Household Science* (New York: n.p., 1853). For textbooks used in women's academies for other sciences see Woody, *History of Women's Education,* vol. 1, app.

40. Phelps's 1834 text for children, *Chemistry for Beginners,* was slightly more successful, and editions continued through the 1860s. Her most popular book was *Familiar Lectures on Botany,* which was reprinted dozens of times and had sold 230,000 copies by 1870. Phelps's biographer Emma Lydia Bolzau has attributed the failure of the chemistry texts to their derivative nature; see Bolzau, *Almira Hart Lincoln Phelps* (Lancaster, Penn.: Science Press, 1936): 235–36.

41. Phelps, *Chemistry for Beginners* (1867), 11. Marcet's most sustained discussion of chemistry and religious faith appeared in the closing paragraph of her book: "To God alone man owes the admirable faculties which enable him to improve and modify the productions of nature. . . . In contemplating the works of the creation, or studying the inventions of art, let us, therefore, never forget the Divine source from which they proceed; and thus every acquisition of knowledge will prove a lesson of piety and virtue" (1822 edition, [e.g.], 327).

42. Phelps, *Chemistry for Beginners* (1839), 5 (bread making); (1867), 11 (housekeeping).
43. M. J. B. Orfila, *Practical Chemist*, trans. from the French by John Coxe (Philadelphia: n.p., 1818); William Henry, *The Elements of Experimental Chemistry* (Philadelphia: n.p., 1822–1823); John Gorham, *The Elements of Chemical Science* (Boston: n.p., 1819); Comstock, *Elements of Chemistry* (see n. 11); J. Dorman Steele, *Fourteen Weeks in Chemistry* (New York: n.p., 1867); and Jeremiah Joyce, *Dialogues in Chemistry* (New York: n.p., 1818). The quotation is from the third London edition, with additional notes by an American professor of chemistry. In her third London edition Marcet stated that her format (a teacher and two students) was borrowed from a book titled *Scientific Dialogues*. This was probably an earlier book by Joyce, who also wrote *Dialogues on the Microscope*. See Marcet's preface, 1819.
44. Amos Eaton, *Chemical Instructor* (Albany: n.p., 1822), dedication. The work went into four editions in Albany (1822, 1826, 1828, 1833). On the relationship between Eaton, Phelps, and Willard see Lois Barber Arnold, *Four Lives in Science: Women's Education in the Nineteenth Century* (New York: Schocken, 1984).
45. Eaton, *Chemical Instructor* (1822), title page.
46. As early as 1809, Marcet's New Haven publishers added a description and plate of the pneumatic cistern of Yale College, a short account of artificial mineral waters, and an appendix consisting of treatises on dyeing, tanning, and currying; see 1819 New Haven edition.
47. The Boston Girls High School has been credited with being the first school to offer the teaching of chemistry with laboratory instruction, in 1865. By 1871 many high schools had chemistry laboratories. See Sidney Rosen, "The Rise of High School Chemistry in America (to 1920)," *Journal of Chemical Education* 33 (1956): 627–33, on 628.
48. Warner, "Science Education for Women," 65–66. See Margaret Rossiter, *Women Scientists in America: Struggles and Strategies to 1940* (Baltimore: Johns Hopkins University Press, 1982), for a full discussion of this emergence of women scientists in the mid- and late nineteenth century.

3. From Individual Practitioner to Regular Physician

1. "Academy of Medicine of Cincinnati, Abstract of Proceedings and Papers, May 4, 1857," *Cincinnati Medical Observer* 2 (1857): 458.
2. On this point, see, for instance, Bruce Sinclair, *Philadelphia's Philosopher Mechanics: History of the Franklin Institute, 1824–1865* (Baltimore: Johns Hopkins University Press, 1974), 241 ff.; Ian R. Tyrrell, *Sobering Up: From Temperance to Prohibition in Antebellum America, 1800–1860* (Westport, Conn.: Greenwood Press, 1979), 197; Frederick Merk and Lois Bannister Merk, *Manifest Destiny and Mission in American History: A Reinterpretation* (New York: Knopf, 1963), passim; Wilfred E. Binkley, *American Political Parties: Their Natural History*, 2d ed. (New York: Knopf, 1943), 181–234; Richard Hofstadter, *The Idea of a Party System: The Rise of Legitimate Opposition in the United States, 1780–1840* (Berkeley and Los Angeles: University of California Press, 1969), 247 ff.; George H. Daniels, *American Science in the Age of Jackson* (New York: Columbia University Press, 1968), 3–33; Sally Gregory Kohlstedt, *The Formation of the American Scientific Community: The American Association for the Advancement of Science, 1848–1860* (Urbana: Univer-

sity of Illinois Press, 1976), passim; Louis Hartz, *Economic Policy and Democratic Thought, Pennsylvania, 1776–1861* (Chicago: Quadrangle Press, 1968), 181 ff.; Oscar and Mary Flugg Handlin, *Commonwealth. A Study of the Role of Government in the American Economy: Massachusetts, 1774–1861* rev. ed. (Cambridge: Harvard University Press, 1969), 203–44; Alfred D. Chandler, Jr., *The Visible Hand: The Managerial Revolution in American Business* (Cambridge: Harvard University Press, 1977), 79 ff.; David A. Hounshell, *From the American System to Mass Production, 1800–1932: The Development of Manufacturing Technology in the United States* (Baltimore: Johns Hopkins University Press, 1984), 15–66; Timothy L. Smith, *Revivalism and Social Reform: American Protestantism on the Eve of the Civil War* (Nashville: Abingdon Press, 1957), 15–33; Alan I Marcus, "Am I My Brother's Keeper: Reform Judaism in the American West, Cincinnati, 1840–1870," *Queen City Heritage* 44 (1986): 3–19; Marcus, "National History From Local: Social Evils and the Origin of Municipal Services in Cincinnati," *American Studies* 22, no. 2 (Fall 1981): 23–39; and Marcus, *Plague of Strangers: Social Groups and the Origin of City Services in Cincinnati, 1819–1870* (Columbus: Ohio State University Press, 1991).

3. Daniel Drake, "Voluntary Medical Societies in Kentucky," *Western Journal of the Medical and Physical Sciences* 11 (1837–38): 340; and Drake, "On Purposes of Scientific Societies," address delivered to the Lexington Medical Society, November 14, 1823 (manuscript housed at the History of Health Sciences Library and Museum, Health Sciences Library, University of Cincinnati Medical Center).

4. For a brief history of some early societies, see Otto Juettner, *Daniel Drake and His Followers* (Cincinnati: Harvey, 1909), 437–40; and Daniel Drake, "Cincinnati Medical Society" and "Repeal of the Medical Law of Ohio, Qualified Approval. Suggestions About Another Law," *Western Journal of the Medical and Physical Sciences* 7 (1833–34): 476–83; John P. Harrison, *Address Delivered to the Hamilton County Medical Association, January 6, 1838,* (Cincinnati: J. Moore and Repp, 1838); and Daniel Drake, "Formation of Professional Character; Introductory Lecture to the First Session of the Medical Department of Cincinnati College, November 2, 1835" (manuscript housed at the History of Health Sciences Library and Museum), 16. Medical politics also played a part in early medical societies. Individuals often used them to further their own medical political ambitions. Drake was particularly adroit at this practice. See Emmet F. Horine, *Daniel Drake (1752–1852): Pioneer Physician to the Midwest* (Philadelphia: University of Pennsylvania Press, 1961), 166–67, 289–301; Joseph F. Kett, *The Formation of the American Medical Profession: The Role of Institutions, 1780–1860* (New Haven: Yale University Press, 1968), 79–94. Also see Daniel Drake, comp., "Documents Relative to the Medical College of Ohio," *Western Journal of Medical and Physical Sciences* 6 (1832–33): 632–40 and "The Medical College of Ohio," 7 (1833–34): 623–51.

5. Drake, "On Scientific Societies," 9. For discussions of the process of early nineteenth-century medical science, see Daniels, *American Science in the Age of Jackson,* 63–117, and Henry D. Shapiro, "Daniel Drake: The Scientist as Citizen," in Henry D. Shapiro and Zane L. Miller, eds., *Physician to the West: Selected Writings of Daniel Drake on Science and Society* (Lexington: University Press of Kentucky, 1970), xi–xxii.

6. Rush Welter, *The Mind of America, 1820–1860* (New York: Columbia University Press, 1975), 142–57. Welter politicizes the notion of success and marks out a

conservative and a democratic view. Also of use on this point are Fred Somkin, *Unquiet Eagle: Memory and Desire in the Idea of American Freedom, 1815–1860* (Ithaca: Cornell University Press, 1967), 11–48; John William Ward, *Andrew Jackson: Symbol for an Age* (New York: Oxford University Press, 1955), 166–80; Stow Persons, *The Decline of American Gentility* (New York: Columbia University Press, 1973), 51–71; Charles C. Cole, Jr., *The Social Ideas of Northern Evangelists, 1826–1860* (New York: Columbia University Press, 1954), 96–131, 165–91; and John R. Bodo, *The Protestant Clergy and Public Issues, 1812–1848* (Princeton: Princeton University Press, 1954), 152–91.

7. See, for instance, Lee Benson, *The Concept of Jacksonian Democracy* (Princeton: Princeton University Press, 1961), passim; Marvin Meyers, *The Jacksonian Persuasion: Politics and Belief* (Stanford: Stanford University Press, 1957), 3–33; and Welter, *Mind*, 77–104. For secrecy in early nineteenth-century medical societies, see both Lee D. Van Antwerp, "Kappa Lambda, Elf or Ogre?" *Bulletin of the History of Medicine* 17 (1945): 327–50; and Phillip Van Ingen, "Remarks on 'Kappa Lambda, Elf or Ogre?' and a Little More Concerning the Society," *Bulletin of the History of Medicine* 18 (1945): 513–38.

8. Samuel Hanbury Smith, *An Introductory Lecture, delivered in the Cincinnati Medical Institute, April 3, 1848* (Cincinnati: Robinson and Jones, 1848), 9. Also see Kett, *Formation*, 100–32; and James Harvey Young, *The Toadstool Millionaires: A Social History of Patent Medicines in America Before Federal Regulation* (Princeton: Princeton University Press, 1961), 44–57. For attacks on monopolies in other spheres, see, for example, A. H. Chroust, *The Rise of the Legal Profession in America*, vol. 1 (Norman: University of Oklahoma Press, 1965), 137–69; and Bray Hammond, "Banking in the Early West: Monopoly, Prohibition and Laissez Faire," *Journal of Economic History* 8 (1948): 1–25.

9. See, for example, Daniel Drake, "[Review of Samuel Thomson's The People's Doctor]," *Western Journal of the Medical and Physical Sciences* 3 (1829): 393–420, 455–62. For the situation throughout America generally, see Madge E. Pickard and R. Carlyle Buley, *The Midwest Pioneer: His Ills, Cures and Doctors* (Crawfordsville: R. E. Banta, 1945), 167–239; Martin Kaufman, *Homeopathy in America: The Rise of a Medical Heresy* (Baltimore: Johns Hopkins University Press, 1971), 23–30; Kett, *Formation*, 135–37; and John F. Gray, *Early Annals of Homeopathy in New York* (New York: n.p., 1863), 9–20.

10. Of the physicians listed as living in Cincinnati and licensed to practice in Ohio in *The Cincinnati Directory, for the Year 1831: Containing the Names of the Inhabitants, Their Occupations, Place of Business and Dwelling-Houses* (Cincinnati: Robinson and Fairbanks, 1831), none are known to have become irregulars. This was the last official listing of licensed practitioners before the Ohio licensing law was replaced. No work adequately traces the change from individual medical practitioner to regular physician. Useful are Kett, *Formation*, 156–60; and Daniel H. Calhoun, *Professional Lives in America, Structure and Aspirations* (Cambridge: Harvard University Press, 1965), 20–58. The historiography of eclectics is somewhat better. See Alex Berman, "The Impact of the Nineteenth Century Botanico-Medical Movement on American Pharmacy and Medicine" (Ph.D. diss., University of Wisconsin, 1954, chapter 6, and Berman, "Neo-Thomsonianism in the United States," *Journal of the History of Medicine and Allied Sciences* 11 (1956): 133–55.

11. John Bell, *An Introductory Lecture, Delivered to the Students of the Medical College*

of Ohio, November 4, 1850 (Cincinnati: Daily Commercial, 1850), 15. The classic statement about the process of science as practiced by irregulars is Oliver Wendell Holmes, "Homeopathy and Its Kindred Delusions," read before the Boston Society for the Diffusion of Useful Knowledge, 1842, printed in Oliver Wendell Holmes, *Medical Essays, 1842–1882* (Boston: Houghton, Mifflin, 1893), 2–102.

12. George Mendenhall, "Sanitary Survey," *Western Lancet* 11 (1850): 679; and Mendenhall, "Public Health," *Cincinnati Medical Observer* 1 (1856): 456.

13. John P. Harrison, *On the Reciprocal Obligations of the Medical Profession and Society: An Introductory Lecture, Delivered at the Medical College of Ohio, November 11, 1846* (Cincinnati: R. P. Donough, 1846), 18; and Cornelius G. Comegys, *The Discouragements and Encouragements of the Medical Students and a Proposition for the Legal Protection of the Medical Profession* (Cincinnati: Achilles Pugh, 1856), 4–5.

14. Joseph R. Buchanan, "Medical Reform," *Eclectic Medical Journal* 8 (1849): 178–83. For works that discuss the attack on regulars as well as their response in the United States generally, see Harris L. Coulter, "Political and Social Aspects of Nineteenth Century Medicine in the United States: The Formation of the American Medical Association and Its Struggle with Homeopathic and Eclectic Physicians," (Ph.D. diss., Columbia University, 1970), passim; Kaufman, *Homeopathy,* chapters 3–6; Thomas N. Bonner, *Medicine in Chicago, 1850–1950: A Chapter in the Social and Scientific Development of a City* (New York: American Book–Stratford Press, 1957), 40–42; and William G. Rothstein, *American Physicians in the Nineteenth Century* (Baltimore: Johns Hopkins University Press, 1972), 165–73.

15. For the charge of popishness, see Ray Billington, *The Protestant Crusade, 1800–1860: A Study of the Origins of American Nativism* (Chicago: Quadrangle Books, 1964), 193–219, 345–79, passim.

16. Trustees of Cincinnati Township, *Annual Report Concerning the Commercial Hospital of Cincinnati* (Cincinnati: n.p., 1849), 9–14; State of Ohio, "An Act to Establish a Commercial Hospital and Lunatic Asylum for the State of Ohio," *Statutes* 19 (1821): 58–66; and Thomas V. Morrow, "Memorial of the Eclectic Medical Society of Cincinnati, to the Legislature of Ohio," *Eclectic Medical Journal* 8 (1849): 130–41.

17. See, for example, Thomas V. Morrow, "The Hospital Bill," *Eclectic Medical Journal* 8 (1849): 142–44; Morrow, "Memorial," 130–37; *Memorial of B. L. Hill, M. D., of Cincinnati, to the Arguments of the Faculty of the Medical College of Ohio* (Columbus: n.p., 1849), passim; Leonidas M. Lawson, "The Quacks and the Commercial Hospital," *Western Lancet* 9 (1849): 262–70. See also John F. Beaver, "Report of the Select Committee of House Bill No. 118," *Appendix to the Journal of the Senate of the Ohio General Assembly* 47 (1849): 336–53; William Bennett, "Report of the Standing Committee on Medical Colleges and Societies, on House Bill No. 118," *Appendix,* 161–74; *Journal of the Senate of the Ohio General Assembly* 47 (1849): 518; *Journal of the House of the Ohio General Assembly* 47 (1849): 269 and 346–49; Leonidas M. Lawson, "Physicians in the Legislature," *Western Lancet* 9 (1849): 271–72; and George Mendenhall, "The Commercial Hospital and the Medical College of Ohio," *Western Lancet* 11 (1850): 268–70.

18. See, for example, Joseph R. Buchanan, "The Ohio Medical College—Wesleyan University and the Commercial Hospital," *Eclectic Medical Journal* 11 (1851): 275–

80 and, in the same journal, his "Medical College of Ohio," 13 (1853): 422–24, and "The Ohio Medical College and the Commercial Hospital of Ohio," 18 (1858): 243–45.

19. *Cincinnati Daily Gazette,* May 1, 1849, p. 2; May 3, 1849, p. 2; May 9, 1849, p. 2; and May 11, 1849, p. 2. The irregulars' claims and the regulars' counterclaims set off a pamphlet war, extending beyond the epidemic's conclusion. See, for example, S. A. Latta, *The Cholera in Cincinnati; Or a Connected View of the Controversy Between the Homeopathists and the Methodist Expositor* (Cincinnati: Morgan and Overend, 1850); and Adam Miller, *Review of Dr. S. A. Latta's Pamphlet, Entitled "The Cholera in Cincinnati; Or a Connected View of the Controversy Between the Homeopathists and the Methodist Expositor* (Cincinnati: Ben Franklin Printing, 1850). Medical journals also took up the fight. See, for instance, Leonidas M. Lawson, "Cholera Reports and Statistics," *Western Lancet* 10 (1849): 59–61 and "Quacks in Cincinnati," *Western Lancet* 10 (1849): 196–98; Thomas C. Carroll, "Observations on the Asiatic Cholera, as It Appeared in Cincinnati During the Years of 1849 and 1850," *Western Lancet* 15 (1854): 340–46; Joseph R. Buchanan, "Medical Politics in Cincinnati, Etc.," *Eclectic Medical Journal* 8 (1849): 375–84.

20. *Cincinnati Daily Atlas,* May 18, 1849, p. 2; *Cincinnati Daily Gazette,* May 12, 1849, p. 2; May 16, 1849, p. 2; May 19, 1849, p. 2; May 21, 1849, p. 2; and May 25, 1849, p. 2; Joseph R. Buchanan, "Another Medical Humbug," *Eclectic Medical Journal* 8 (1849): 264–65 and "To the City Council and Board of Health of Cincinnati," *Eclectic Medical Journal* 8 (1849): 265–66; and James Taylor, "Church and State," *Physio-Medical Recorder* 17 (1849): 175–76.

21. *City Council Minutes,* May 18, 1849 and May 28, 1849; *Cincinnati Daily Gazette,* May 21, 1849, p. 2; and "Board of Health—Proceedings of City Council," *Eclectic Medical Journal* 8 (1849): 266–70.

22. *City Council Minutes,* May 28, 1849; *Cincinnati Daily Commercial,* May 23, 1849, p. 2; May 28, 1849, p. 2; and May 30, 1849, p. 2; and Joseph R. Buchanan, "Cholera in Cincinnati—Triumph of Liberal Principles," *Eclectic Medical Journal* 8 (1849): 284–88.

23. *Cincinnati Daily Commercial,* May 30, 1849, p. 2; *Cincinnati Daily Atlas,* May 31, 1849, p. 3; and Lawson, "Cholera Report," 60–61.

24. For Cincinnati, see, for example, Leonidas M. Lawson, "Medical Reform," *Western Lancet* 5 (1846–47): 67–73; Lawson, "Present Condition and Future Prospects of the Medical Profession," *Western Lancet* 11 (1850): 59–62; Thomas Wood, "Professional Follies and Professional Feuds," *Western Lancet* 14 (1853): 447–50; and "Pugnacity of the Medical Profession," *Western Lancet* 16 (1855): 443–45. For the United States generally, see Kaufman, *Homeopathy,* 51–55; Rothstein, *American Physicians,* 101–14; and John Duffy, *The Healers: The Rise of the Medical Establishment* (New York: McGraw-Hill, 1976), 183–88.

25. Leonidas M. Lawson, "Legal Regulation of Medicine and Surgery," *Western Lancet* 2 (1843–44): 437; and Smith, *Introductory Lecture,* 3 and 9.

26. Benjamin Franklin Richardson, "The State of the Profession," *Western Lancet* 18 (1857): 317; and John A. Murphy, *An Introductory Lecture, Delivered to the Students of the Miami Medical College, October 30, 1854* (Cincinnati: T. Wrightson, 1855), 8.

27. John P. Harrison, *On the Benefits Accruing to Society From the Medical Profession: An Introductory Lecture, Delivered to the Students of the Medical College of Ohio,*

November 8, 1843, (Cincinnati: R. P. Donough, 1843), 4 and Harrison, *On Reciprocal Obligations,* 5.

28. *Constitution and By-Laws of the Medico-Chirurgical Society of Cincinnati* (Cincinnati: Chronicle and Atlas, 1850), 3, 6, and 8; *Cincinnati Daily Gazette,* December 8, 1852, p. 2; J. P. Walker, "Annual Report of the Secretary of the Cincinnati Medical Society," *Western Lancet* 15 (1854): 20–21; and J. R. Atkins, "Monthly Report of the Cincinnati Medical Society," *Western Lancet* 17 (1856): 90.

29. Leonidas M. Lawson, "New Medical Schools," *Western Lancet* 13 (1852): 710–11; *Annual Announcement of Lectures of the Medical College of Ohio for the Session of 1852–53* (Cincinnati: T. Wrightson, 1852), 9–11; *Annual Announcement of the Cincinnati College of Medicine and Surgery for the Session of 1852–53* (Cincinnati: T. Wrightson, 1852), 6, 11; *Third Annual Announcement of Lectures of the Miami Medical College of Cincinnati for the Session of 1854–55* (Cincinnati: T. Wrightson, 1854), 4–6; *Annual Announcement of the Cincinnati College of Medicine and Surgery for the Session 1855–56* (Cincinnati: T. Wrightson, 1855), 13; *Fifth Annual Announcement of Lectures of the Miami Medical College for the Session of 1856–57* (Cincinnati: Wrightson, 1856), 1–2, 5–6; and *Thirty-Seventh Annual Announcement of Lectures of the Medical College of Ohio for the Session of 1856–57* (Cincinnati: Wrightson, 1856), 7–9, 11–13.

30. Benjamin Franklin Richardson, "Medical Societies," *Western Lancet* 18 (1857): 697–700.

31. McIllvaine is quoted in Thomas Wood, "[Editorial]," *Western Lancet* 18 (1857): 66 and quoted retrospectively in *Cincinnati Daily Gazette,* February 7, 1860, p. 2. Also see A. H. Baker, "Cod-Liver Oil Not a Medicine," *Cincinnati Medical and Surgical News* 1 (1860): 63; and James A. Thacker, "Proceedings of the Cincinnati Academy of Medicine, February 5, 1860," *Cincinnati Lancet and Observer* 3 (1860): 160–61. For medical organizations in other cities established for similar purposes, see, for instance, "Proceedings of a Meeting for the Establishment of an Academy of Medicine and Surgery," *New York Journal of Medicine* 8 (1847): 125–27; Thomas N. Bonner, *The Kansas Doctor: A Century of Pioneering* (Lawrence: University of Kansas Press, 1959), 41–43 and Bonner, *Medicine in Chicago,* 69–74.

32. J. A. Thacker, *Brief Historical Sketch and Constitution and By-Laws of the Academy of Medicine of Cincinnati, Ohio* (Cincinnati: P. C. Browne, 1865), 5; E. B. Stevens, "Medico-Chirurgical Society," *Cincinnati Medical Observer* 2 (1857): 90 and "Cincinnati Academy of Medicine," 137 and 186; and *Cincinnati Daily Gazette,* February 9, 1857, p. 2.

33. Thacker, *Brief Sketch,* 7–9; C. B. Hughes, "Proceedings of the Cincinnati Academy of Medicine, March 8, 1858," *Cincinnati Lancet and Observer* 1 (1858): 238; and A. M. Johnson, "Proceedings of the Cincinnati Academy of Medicine, June 7, 1858," *Cincinnati Lancet and Observer* 1 (1858): 417–18. The academy's minutes prior to 1869 no longer exist. For a similarly organized medical society in a different city, see "Constitution and By-Laws of the New York Academy of Medicine," *New York Journal of Medicine* (1847): 255–57.

34. Stevens was quoted in A. M. Johnson, "Proceedings of the Cincinnati Academy of Medicine, April 5, 1858," *Cincinnati Lancet and Observer* 1 (1858): 300–1. On these points, see further articles in the same journal, for instance, A. M. Johnson, "Proceedings of the Cincinnati Academy of Medicine, October 10, 1858," 1 (1858): 699–80; W. H. Reynolds, "Diphtheria," 2 (1859): 265–69; J. A. Thacker,

"Proceedings of the Cincinnati Academy of Medicine, July 11, 1859," 2 (1859): 475–83; J. A. Murphy, "The Cincinnati Academy of Medicine," 3 (1860): 94; and see John Keith Moore, "Discussion in the Academy of Medicine, November 12, 1860," *Cincinnati Medical and Surgical News* 1 (1860): 372–75.

35. C. B. Hughes, "Academy of Medicine, of Cincinnati, Abstract of Proceedings and Papers, Regular Meeting May 4th, 1857," *Cincinnati Medical Observer* 2 (1857): 457–58.

36. See, for example, C. B. Hughes, "Academy of Medicine of Cincinnati, Abstract of Proceedings and Papers, June 1, 1857," *Cincinnati Medical Observer* 2 (1857): 459; A. M. Johnson, "Proceedings of the Cincinnati Academy of Medicine, August 9, 1858," *Cincinnati Lancet and Observer* 1 (1858): 556–62; W. T. Brown, "Milk and Its Adulterations," *Cincinnati Lancet and Observer* 1 (1858): 577–87; *Cincinnati Daily Enquirer,* September 7, 1858, p. 3; and J. A. Thacker, "Proceedings of the Cincinnati Academy of Medicine, September 5, 1859," *Cincinnati Lancet and Observer* 2 (1859): 597–607. Also see Alan I Marcus, *Plague of Strangers,* 165–67, 171–75, 198–200.

37. Murphy is quoted in W. T. Brown, "Proceedings of the Cincinnati Academy of Medicine, March 2, 1863," *Cincinnati Lancet and Observer* 6 (1863): 219.

4. Diagnosing Unnatural Motherhood

1. J. M. Carr, "Puerperal Insanity," *Cincinnati Lancet-Clinic* 7 (1881): 537–42.

2. I do not mean to imply that studies of the history of ideas about insanity are inappropriate. Such studies as Norman Dain, *Concepts of Insanity in the United States, 1789–1865* (New Brunswick: Rutgers University Press, 1964) and Ellen Dwyer, "A Historical Perspective," in Cathy Spatz Widom, ed., *Sex Roles and Psychopathology* (Bloomington: Indiana University Press, 1984), 19–48 are absolutely essential to our understanding of the history of insanity. In her essay, Dwyer goes beyond a history of ideas approach and attempts to test how ideas about gender and insanity related to actual practice. This relationship between ideas and practice deserves more attention. In another study, Dwyer is particularly interested in comparing ideas and practice: Ellen Dwyer, "The Weaker Vessel: Legal Versus Social Reality in Mental Commitment in Nineteenth-Century New York," in D. Kelly Weisberg, ed., *Women and the Law: A Social Historical Perspective,* vol. 2 (Cambridge, Mass.: Schenkman, 1982), 85–106. Also in that volume is a twentieth-century study that compares ideas and practice: Robert T. Roth and Judith Lemer, "Sex-based Discrimination in the Mental Institutionalization of Women," 107–39.

3. I do not think the approach to insanity taken by Mark S. Micale, "On the Disappearance of Hysteria: A Medical and Historical Perspective" (paper read at the annual meeting of the American Association for the History of Medicine, Birmingham, April 1989) is helpful. Micale explains the disappearance of hysteria as due to more specific medical definitions. The real question, however, is how and why definitions change.

4. Michel Foucault, *Madness and Civilization: A History of Insanity in the Age of Reason,* trans. Richard Howard (New York: Pantheon, 1965).

5. For discussions of the labeling theory, see Roy Porter, *A Social History of Madness* (New York: New American Library, 1987); Agnes Miles, *The Neurotic Woman: The Role of Gender in Psychiatric Illness* (New York: New York University Press,

1988); Elaine Showalter, "Victorian Women and Insanity," in Andrew T. Scull, ed., *Madhouses, Mad-Doctors, and Madmen: The Social History of Psychiatry in the Victorian Era* (Philadelphia: University of Pennsylvania Press, 1981), 313–31; Richard W. Fox, *So Far Disordered in Mind: Insanity in California, 1870–1930* (Berkeley and Los Angeles: University of California Press, 1978); Nancy E. Waxler, "Culture and Mental Illness: A Social Labeling Perspective," *Journal of Nervous and Mental Disease* 159 (1974): 379–95.

6. See, for example, Bonnie Ellen Blustein, "A Hollow Square of Psychological Science: American Neurologists and Psychiatrists in Conflict," in Scull, "Madhouses," 241–70, and in the same volume Barbara Sicherman, "The Paradox of Prudence: Mental Health in the Gilded Age," 218–40.

7. For example, Waxler, "Culture and Mental Illness"; Showalter, "Victorian Women and Insanity," and *The Female Malady: Women, Madness, and English Culture, 1830–1980* (New York: Pantheon, 1987); Miles, *The Neurotic Woman;* Dwyer, "The Weaker Vessel."

8. Carroll Smith-Rosenberg, "The Hysterical Woman: Sex Roles and Role Conflict in Nineteenth-Century America," *Social Research* 39 (1972): 562–83; Joan Jacobs Brumberg, *Fasting Girls: The Emergence of Anorexia Nervosa as a Modern Disease* (Cambridge, Mass.: Harvard University Press, 1988).

9. For an explanation of "illness" and "disease," see Arthur Kleinman, *Social Origins of Distress and Disease: Depression, Neurasthenia, and Pain in Modern China* (New Haven: Yale University Press, 1986); Claudine Herzlich and Janine Pierret, *Illness and Self in Society,* trans. Elborg Forster (Baltimore: Johns Hopkins University Press, 1987); Arthur Kleinman, *Patients and Healers in the Context of Culture* (Berkeley and Los Angeles: University of California Press, 1980); Peter Conrad and Rochelle Kem, eds., *The Sociology of Health and Illness: Critical Perspectives* (New York: St. Martin's, 1981); Peter Wright and Andrew Treacher, eds., *The Problem of Medical Knowledge: Examining the Social Construction of Medicine* (Edinburgh: Edinburgh University Press, 1982), especially essays by David Ingleby, "The Social Construction of Mental Illness," 123–41, and Karl Figlio, "How Does Illness Mediate Social Relations: Workmen's Compensation and Medico-Legal Practices, 1890–1940," 174–224; Bryan S. Turner, *The Body and Society Explorations in Social Theory* (Oxford: Basil Blackwell, 1984).

10. Milton H. Hardy "Puerperal Insanity," *Western Medical Review* 3 (1898): 14.

11. W. H. B. Stoddart, "A Clinical Lecture on Insanity in Relation to the Childbearing State and the Puerperium," *The Clinical Journal* 14 (1899): 242; Hardy, "Puerperal Insanity," 14; R. M. Wigginton, "Puerperal Insanity," *Transactions of the Wisconsin State Medical Society* 60 (1975): 40; L. R. Landfear, "Puerperal Insanity," *Cincinnati Lancet and Observer* 19 (1876): 54; Harry L. K. Shaw, "A Case of Insanity of Gestation," *Albany Medical Annals* 19 (1898): 459–62; Fleetwood Churchill, "On the Mental Disorders of Pregnancy and Childbed," *American Journal of Insanity* 7 (1850–51): 297–317.

12. W. F. Menzies, "Puerperal Insanity: An Analysis of One Hundred and Forty Consecutive Cases," *American Journal of Insanity* 50 (1893–94): 147–85; George H. Rohe, "Lactational Insanity," *Journal of the American Medical Association* 21 (1893): 325–27; W. G. Stearns, "The Psychiatric Aspects of Pregnancy," *Obstetrics* 3 (1901): 23–26, 32–36; Wigginton, "Puerperal Insanity," 41.

13. These symptoms were listed in numerous nineteenth-century articles. Since we

will be dealing with the symptoms in more detail later, I will refrain from citing all of the articles here.

14. Landfear, "Puerperal Insanity," 54; Wigginton, "Puerperal Insanity," 41. See also R. E. Haughton, "Puerperal Mania," *Cincinnati Lancet and Observer* 9 (1866): 713; B.C. Hirst, "Six Cases of Puerperal Insanity," *Journal of the American Medical Association* 12 (1889): 29; F. C. Femald, "Puerperal Insanity," *American Journal of Obstetrics* 20 (1887): 714; George Byrd Harrison, "Puerperal Insanity," *American Journal of Obstetrics* 30 (1894): 530. Arthur C. Jelly, "Puerperal Insanity," *Boston Medical and Surgical Journal* 144 (1901): 271 is the only medical writer I found who claimed that puerperal insanity was not common.

15. Hirst, "Six Cases of Puerperal Insanity," 29; Landfear, "Puerperal Insanity," 57; Anna Burnet, "Puerperal Insanity: Cause, Symptoms and Treatment," *Woman's Medical Journal* 9 (1899): 269; "Abstracts," *American Journal of Obstetrics* 13 (1880): 641; T. W. Fisher, "Two Cases of Puerperal Insanity," *Boston Medical and Surgical Journal* 79 (1869): 233–34; Charles E. Ware, "A Case of Puerperal Mania," *American Journal of Medical Sciences* 26 (1853): 346; Churchill, "On the Mental Disorders of Pregnancy and Childbed," 309.

16. On the generally good prognosis expected of puerperal insanity patients, see Stearns, "The Psychiatric Aspects of Pregnancy," 24; William Mercer Sprigg, "Puerperal Insanity: Prognosis and Treatment," *American Journal of Obstetrics* 30 (1894): 537. On home versus hospital treatment, see Jelly, "Puerperal Insanity," 275; Wigginton, "Puerperal Insanity," 43; Landfear "Puerperal Insanity," 59; W. W. Godding, "Puerperal Insanity," *Boston Medical and Surgical Journal* 91 (1874): 318–19.

17. On bleeding as a treatment for puerperal insanity, see Sprigg, "Puerperal Insanity: Prognosis and Treatment," 540, Landfear, "Puerperal Insanity," 58; Churchill, "On the Mental Disorders of Pregnancy and Childbed," 315; J. A. Wright, "Puerperal Insanity," *Cincinnati Lancet-Clinic* 23 (1889): 651; Victor H. Coffman, "Puerperal Mania," *Nebraska State Medical Association Proceedings* 4 (1872): 18. On purging as a treatment, see Haughton, "Puerperal Mania," 729; A. Bryant Clarke, "On the Treatment of Puerperal Mania by Veratrum Viride," *Boston Medical and Surgical Journal* 59 (1859): 237–39; J. MacDonald, "Observations on Puerperal Mania," *New York Medical Journal* 1 and 2 (1831): 279; Thomas Lightfood, "Puerperal Mania; Its Nature and Treatment," *Medical Times and Gazette* 21 (1850): 274. Many physicians recommended that the patient be removed from family and friends. For example, see J. A. Reagan, "Puerperal Insanity," *Charlotte Medical Journal* 14 (1899): 309; Femald, "Puerperal Insanity," 720; J. Thompson Dickson, "A Contribution to the Study of the So-Called Puerperal Insanity," *Journal of Mental Science* 16 (1870–71): 390; C. S. May, "Puerperal Insanity, with Statistics Regarding Sixteen Cases," *Proceedings of the Connecticut State Medical Society* 85 (1877): 106; Carr, "Puerperal Insanity," 540; Wigginton, "Puerperal Insanity," 44; Landfear, "Puerperal Insanity," 59. Nearly every physician who wrote of treatment recommended rest, food, and some form of sedation. For examples not yet cited, see W. S. Armstrong, "Case of Puerperal Mania," *Atlanta Medical and Surgical Journal* 8 (1867): 419; Thomas H. Mayo, "Puerperal Mania," *Southern Medical Record* 4 (1874): 84; J. P. Reynolds "Puerperal Mania," *Boston Medical and Surgical Journal* 72 (1855): 281; W. A. McPheeters, "Forceps; Puerperal Mania," *New Orleans Medical and Surgical Journal* 16 (1859): 660; Edward Kane

"Puerperal Insanity," *Medical Independent and Monthly Review of Medicine and Surgery* 2 (1856): 156; Horace Palmer, "A Case of Puerperal Mania," *Cincinnati Lancet and Observer* 2 (1859): 5; Horatio Storer, "Puerperal Mania; Recovery," *Boston Medical and Surgical Journal* 55 (1856–57): 20.

18. See for example the explanation of Elaine Showalter in *The Female Malady,* especially 57–59, 71–72.

19. Wigginton, "Puerperal Insanity," 43.

20. For example, see W. L. Worcester, "Is Puerperal Insanity a Distinct Clinical Form?" *American Journal of Insanity* 47 (1890–91): 56; Edward J. Ill, "A Clinical Contribution to Gynecology," *American Journal of Obstetrics* 16 (1883): 264; Rohe, "Lactational Insanity," 325–26.

21. Allan McLane Hamilton, "Two Cases of Peculiar Mental Trouble Following the Puerperal State," *Boston Medical and Surgical Journal* 94 (1896): 680.

22. Lambert Ott, "Puerperal Mania," *Clinical News,* 1 (1880): 337; W. P. Manton, "Puerperal Hysteria (Insanity?)," *Journal of the American Medical Association* 19 (1892): 61; W. L. Richardson, "Puerperal Septicaemia: Puerperal Mania," *Boston Medical and Surgical Journal* 102 (1880): 448. See also Carr, "Puerperal Insanity," 538; Femald, "Puerperal Insanity," 717; W. D. Haines, "Insanity in the Puerperal State," *Cincinnati Lancet-Clinic* 23 (1889): 371; Worcester, "Is Puerperal Insanity a Distinct Clinical Form?" 55; Burnet, "Puerperal Insanity: Cause, Symptoms and Treatment," 267–69. This is only a sampling of the articles listing these symptoms.

23. Harrison, "Puerperal Insanity," 532; Haines, "Insanity in the Puerperal State," 371; Godding, "Puerperal Insanity," 317; V. H. Taliaferro, "Puerperal Insanity," *Atlanta Medical and Surgical Journal* 15 (1877): 324; Carr, "Puerperal Insanity," 538; Stearns, "The Psychiatric Aspects of Pregnancy," 25. Almost every physician describing symptoms of puerperal insanity listed obscenity. For example, see Stoddart, "A Clinical Lecture," 242; Reagan, "Puerperal Insanity," 308; Menzies, "Puerperal Insanity," 169; Rohe, "Lactational Insanity," 325; Worcester, "Is Puerperal Insanity a Distinct Clinical Form?" 56; May, "Puerperal Insanity," 106; Wigginton, "Puerperal Insanity," 43; MacDonald, "Observations on Puerperal Mania," 268–70; Clarke, "On the Treatment of Puerperal Mania by Veratrum Viride," 238.

24. George H. Rohe, "Some Causes of Insanity in Women," *American Journal of Obstetrics* 34 (1896): 802. See also John Young Brown, "Pelvic Disease in Its Relationship to Insanity in Women," *American Journal of Obstetrics* 30 (1894): 360; Montrose A. Pallen, "Some Suggestions with Regard to the Insanities of Females," *American Journal of Obstetrics* 10 (1877): 207. Carroll Smith-Rosenberg, "Puberty to Menopause: The Cycle of Femininity in Nineteenth-Century America," *Feminist Studies* 1 (1973): 58–72 is a discussion of this attitude.

25. P. V. Carlin, "Insanity of Pregnancy," *Denver Medical Times* 3 (1883–84): 233. See also Churchill, "On the Mental Disorder of Pregnancy and Childbed," 298; C. P. Lee, "Puerperal Mania," *Kansas Medical Index* 2 (1881): 200.

26. Pallen, "Some Suggestions with Regard to the Insanities of Females," 212.

27. For a clear explanation of this point especially with reference to gender, see Dwyer, "A Historical Perspective."

28. C. C. Hersman, "The Relationship between Uterine Disturbance and Some of the Insanities," *Journal of the American Medical Association* 33 (1899): 709; Carr, "Puerperal Insanity," 537. See also Edward Jarvis, "Causes of Insanity," *Boston*

Medical and Surgical Journal 45 (1851): 289–305; Reagan, "Puerperal Insanity," 309; Kane, "Puerperal Insanity."

29. For example see Femald, "Puerperal Insanity," 716; Landfear, "Puerperal Insanity," 56; Jelly, "Puerperal Insanity," 271–72; G. H. Rohe, "The Influence of Parturient Lesions of the Uterus and Vagina, in the Causation of Puerperal Insanity," *Journal of the American Medical Association* 19 (1892): 59–60.

30. About heredity as a factor in puerperal insanity, see Churchill, "On the Mental Disorders of Pregnancy and Childbed," 305; Fisher, "Two Cases of Puerperal Insanity," 233; Dickson, "A Contribution to the Study of the So-called Puerperal Insanity," 382; Taliaferro, "Puerperal Insanity," 328; May, "Puerperal Insanity," 106; Landfear, "Puerperal Insanity," 55; Femald, "Puerperal Insanity," 714; Jelly, "Puerperal Insanity," 272; Reagan, "Puerperal Insanity," 309.

31. This point of view was expressed in many of the articles. See, for example, Carr, "Puerperal Insanity," 537; Jelly, "Puerperal Insanity," 272; Wigginton, "Puerperal Insanity," 40, Femald, "Puerperal Insanity," 715; Wright, "Puerperal Insanity," 648; Harrison, "Puerperal Insanity," 532; Menzies "Puerperal Insanity: An Analysis of One Hundred and Forty Consecutive Cases," 162; Churchill, "On the Mental Disorders of Pregnancy and Childbed," 299–300; Kane, "Puerperal Insanity," 152.

32. Horatio Robinson Storer, "The Medical Management of Insane Women," *Boston Medical and Surgical Journal* 71 (1864): 210–18; Horatio R. Storer, "Cases Illustrative of Obstetric Disease Deductions Concerning Insanity in Women," *Boston Medical and Surgical Journal* 70 (1864): 189–200.

33. Charles A. L. Reed, "The Gynecic Element in Psychiatry with Suggestions for Asylum Reform," *Buffalo Medical and Surgical Journal* 28 (1888–89): 571. See also Charles F. Folsom, "The Prevalence and Causes of Insanity Commitments to Asylums," *Boston Medical and Surgical Journal* 103 (1880): 97–100; J. H. McIntyre, "Disease of the Uterus and Adnexa in Relation to Insanity," *Transactions of the Medical Association of Missouri* (1898): 191–95; Pallen, "Some Suggestions with Regard to the Insanity of Females," 207; W. P. Jones, "Insanity Dependent Upon Physical Disease," *Tennessee State Medical Society Transactions* 47 (1880): 97–104.

34. Femald "Puerperal Insanity," 719.

35. In addition to Reed, other gynecologists calling for specialists in insane asylums included Joseph Wigglesworth, "On Uterine Disease and Insanity," *Journal of Mental Sciences* 30 (1884–85): 509–31; I. S. Stone, "Can the Gynecologist Aid the Alienist in Institutions for the Insane?" *Journal of the American Medical Association* 16 (1891): 870–73; Ernest Hall, "The Gynecological Treatment of the Insanity in Private Practice," *Pacific Medical Journal* 43 (1900): 241–56; Pallen, "Some Suggestions with Regard to the Insanities of Females."

36. Dwyer notes this trend of attributing women's insanity to problems with the female reproductive system in "The Weaker Vessel," "A Historical Perspective," and in her study of two New York asylums, *Homes for the Mad: Life Inside Two Nineteenth-Century Asylums* (New Brunswick: Rutgers University Press, 1987).

37. Physicians who described medical and surgical attention to female asylum inmates include H. A. Tomlinson and Mary E. Bassett, "Association of Pelvic Diseases and Insanity in Women, and the Influence of Treatment on the Local Disease upon the Mental Condition," *Journal of the American Medical Association* 33 (1899): 827, 831; W. P. Manton, "The Frequency of Pelvic Disorders in Insane Women," *American Journal of Obstetrics* 39 (1899): 54–57; Eugene G. Carpenter,

"Pelvic Disease as a Factor of Cause in Insanity of Females and Surgery as a Factory of Cure," *Journal of the American Medical Association* 35 (1900): 545–51; W. J. Williams, "Nervous and Mental Diseases in Relation to Gynecology," *Transactions of the Eighth Annual Meeting of the Western Surgical and Gynecological Association,* Omaha, December 1898 (1899): 49–57; W. O. Henry, "Insanity in Women Associated with Pelvic Diseases," *Annals of Gynecology and Pediatry* 14 (1900–01): 312–20; W. P. Manton "Post Operative Insanity, Especially in Women," *Annals of Gynecology and Pediatry* 10 (1896–97): 714–19; Hall, "The Gynecological Treatment of the Insanity in Private Practice."

38. See, for example, Henry, "Insanity in Women Associated with Pelvic Diseases"; Manton "The Frequency of Pelvic Disorders in Insane Women"; Hall, "The Gynecological Treatment of the Insanity in Private Practice"; Rohe, "The Influence of Parturient Lesions of the Uterus and Vagina, in the Causation of Puerperal Insanity." I do not mean to imply that all doctors were treating puerperal insanity with surgery. Rest and restoration was probably the most popular therapy throughout the century. See Hardy, "Puerperal Insanity."

39. Tomlinson and Bassett, "Association of Pelvic Diseases and Insanity in Women," 827. See also Brown, "Pelvic Disease in Its Relationship to Insanity in Women"; Carpenter, "Pelvic Disease as a Factor of Cause in Insanity of Females and Surgery as a Factory of Cure"; and Williams, "Nervous and Mental Diseases in Relation to Gynecology" for examples of the argument that doctors should only resort to surgery when there is physical disease.

40. Stone, "Can the Gynecologist Aid the Alienist in Institutions for the Insane?"; Manton, "Post-Operative Insanity, Especially in Women"; Clara Barrus, "Gynecological Disorders and Their Relation to Insanity," *American Journal of Insanity* 51 (1894–95): 475–91; Alice May Farnham, "Uterine Disease as a Factor in the Production of Insanity," *Alienist and Neurologist* 8 (1887): 532–47; Adolf Meyer, "On the Diseases of Women as a Cause of Insanity in the Light of Observations of Sixty-Nine Autopsies," *Transactions of the Illinois State Medical Society* (1895): 299–311; C. B. Burr, "The Relation of Gynecology to Psychiatry," *Transactions of the Michigan Medical Society* 18 (1894): 458–64, 478–87. An excellent article about removal of the ovaries in the nineteenth century is Lawrence D. Longo, "The Rise and Fall of Battey's Operation: A Fashion in Surgery," *Bulletin of the History of Medicine* 53 (1979) 244–67.

41. See, for example, Helene Kuhlmann, "A Few Cases of Interest in Gynecology in Relation to Insanity," *State Hospital Bulletin* (New York) 1 (1896): 172–79; Mary D. Jones, "Insanity, Its Causes: Is There in Women a Correlation of the Sexual Function with Insanity and Crime?" *Medical Record* 58 (1900): 925–37; Edward B. Lane, "Puerperal Insanity," *Boston Medical and Surgical Journal* 144 (1901): 606–9; Manton, "Puerperal Hysteria."

42. For example Jones, "Insanity, Its Causes: Is There in Women a Correlation of the Sexual Function with Insanity and Crime?"; Kuhlmann, "A Few Cases of Interest in Gynecology in Relation to Insanity"; Burnet, "Puerperal Insanity: Cause, Symptoms and Treatment"; and the work of Mary Putnam Jacobi. An important study of women physicians working in asylums is Constance M. McGovern, "Doctors or Ladies? Women Physicians in Psychiatric Institutions, 1872–1900," *Bulletin of the History of Medicine* 55 (1981) 88–107.

43. A creative use of case studies to describe the doctor-patient relationship and the doctor-family relationship is Ellen Dwyer, "The Burden of Illness: Families and

Epilepsy," paper read at the annual meeting of the American Association for the History of Medicine, Birmingham, April 1989.

44. McPheeters, "Forceps; Puerperal Mania"; Ware, "A Case of Puerperal Mania"; Lambert, "Puerperal Mania." Numerous case studies included information about cruelty, illness, illegimacy, and stillbirth, although often doctors did not connect the situation to the symptoms. See, for example, W. H. Parish, "Puerperal Insanity," *Transactions of the Obstetrical Society of Philadelphia* 4–7 (1876–79): 50–54.

45. Fisher, "Two Cases of Puerperal Insanity," 233.

46. Harrison, "Puerperal Insanity," 535.

47. Lee, "Puerperal Mania," 205; Lambert, "Puerperal Mania," 337; Denslow Lewis, "Clinical Lecture on Obstetrics and Gynecology Mental and Nervous Derangements in Obstetric Practice," *Clinical Review* 11 (1899–1900): 181.

48. This theme of women denying their husbands and/or expressing fear or hatred of their husbands was common in the case studies.

49. Dickson, "A Contribution to the Study of the So-Called Puerperal Insanity," 383.

50. Storer, "Puerperal Mania Recovery," 20; McDonald, "Observations on Puerperal Mania," 270.

5. The Inventor of the Mustache Cup

1. Brooke Hindle, *Emulation and Invention* (New York: W. W. Norton, 1981), 12–14.

2. On the rather slow emergence of engineers in America and their affinity to large-scale organizations, see Edwin T. Layton, *The Revolt of the Engineers* (Baltimore: Johns Hopkins University Press, 1986), 1–52.

3. On the modern industrial research laboratory, see Leonard S. Reich, *The Making of American Industrial Research: Science and Business at GE and Bell, 1876–1926* (New York: Cambridge University Press, 1985) and George Wise, *Willis R. Whitney, General Electric, and the Origins of U.S. Industrial Research* (New York: Columbia University Press, 1985). For an overview of the modern technological system, including the relative positions of inventors and engineers, see Edwin T. Layton, "Conditions of Technological Development," in Derek J. Price and Ina Spiegel-Rosing, eds., *Science, Technology and Society: A Cross-Disciplinary Perspective* (London: Russell-Sage, 1977), 197–222.

4. See for example Alfred C. Chandler, *The Visible Hand* (Cambridge: Harvard University Press, 1977) on the rise of the modern corporation. On systems building, hierarchy, and systematization see Alan I Marcus and Howard P. Segal, *Technology in America: A Brief History* (San Diego: Harcourt, Brace, and Co., 1989), 140–41, 165–77, 225–43.

5. Alan I Marcus, *Agricultural Science and the Quest for Legitimacy: Farmers, Agricultural Colleges and Experiment Stations, 1870–1890* (Ames: Iowa State University Press, 1985), 4–32.

6. Matthew Josephson, *Edison: A Biography* (New York: Macmillan, 1959), 131–38, 193–99, 281–84, 313–17. Edison's laboratory, though an important industrial one, differed from the later General Electric Laboratory in significant ways. See Robert Friedel and Paul Israel, *Edison's Electric Light: Biography of an Invention* (New Brunswick: Rutgers University Press, 1986), 146–47.

7. George S. Morison, "Address at the Annual Convention," *Transactions of the American Society of Civil Engineers* 33 (June 1895): 467.

8. John F. Kasson, *Civilizing the Machine: Technology and Republican Values in America, 1776–1900* (New York: Grossman, 1977).

9. David Jeremy, "Innovation in American Textile Technology During the Early 19th Century," *Technology and Culture* 14 (January 1973): 40–76.

10. A useful survey of the origins of the New England hydraulic engineering tradition is provided by Charles W. Sherman, "Great Hydraulic Engineers of New England's Classic Period," *Engineering News-Record* 107 (September 24, 1931): 475–79. Any serious study of Storrow should start with the biographical material on him collected as part of the Essex Company Papers at the Merrimack Valley Textile Museum (see Charles Storrow, Biographical File, Essex Company Papers, at the Merrimack Valley Textile Museum, North Andover, Massachusetts).

11. Charles S. Storrow, ed. and trans., *A Treatise on Water Works, for Conveying and Distributing Supplies of Water, with Tables and Examples by Charles S. Storrow* (Boston: n.p., 1835). J. F. D'Aubuisson de Voisins, *A Treatise on Hydraulic, for the Use of Engineers,* trans. Joseph Bennett (Boston: n.p., 1852).

12. Edwin T. Layton, "European Origins of the American Engineering Style of the Nineteenth Century," in Nathan Reingold and Marc Rothenberg, eds., *Scientific Colonialism* (Washington: Smithsonian Press, 1987), 159–64.

13. I have discussed Boyden and Francis and their style of engineering in comparison with both European engineering and the work of American millwrights in Edwin T. Layton, "Scientific Technology, 1845–1900: The Hydraulic Turbine and the Origins of American Industrial Research," *Technology and Culture* 20 (January 1979): 64–89, and Edwin T. Layton, "Millwrights and Engineers: Science, Social Roles and the Evolution of the Turbine in America," in Wolfgang Krohn, Edwin T. Layton, and Peter Weingart, eds., *The Dynamics of Science and Technology, Sociology of the Sciences,* vol. 2 (Dordrecht, Holland: D. Reidel, 1978): 61–87. It should be emphasized that there was among millwrights and other craftsmen a creative elite who sought to base technological practice upon science, though often a vernacular science. James Emerson was not one of these, and he continued to deny that science played any role in turbine improvements, even when the leading designers told him clearly that they did use scientific methods and principles (see below).

14. Francis, *Lowell Hydraulic Experiments,* 71–145.

15. Layton, "Scientific Technology," 81–84.

16. James Emerson, *Report of Water-Wheel Tests at Lowell* (Lowell: n.p., 1872). He considered this the first edition of his book, though it was a preliminary pamphlet of only sixteen pages. The "second edition" added seventy-two pages to the preliminary pamphlet and was titled *Treatise on the Manner of Testing Water-Wheels and Machinery* (Lowell: n.p., 1872). In fact these two were bound together, constituting a single volume of eighty-eight pages. The third edition (more properly the second) was greatly expanded (to 359 pages). It was titled *Treatise Relative to the Testing of Water-Wheels and Machinery, with Various Other Matters Pertaining to Hydrodynamics* (Springfield: n.p., 1881). The fourth edition, which was completed after his removal as director of the Holyoke testing flume, was greatly expanded (to 480 pages) and included greater and more detailed coverage of his social and political philosophies. He published it with a revised subtitle as *Treatise Relative to the Testing of Water-Wheels and Machinery, also of Inventions, Studies, and Experiments, with Suggestions from a Life's Experiences*

(Springfield: n.p., 1892). Emerson published a fifth and a sixth edition (the sixth edition appears to have been a reprint of the fifth, and properly they constitute one edition). They were further expanded in the sixth edition (to 569 pages) and bore the same title as the fourth edition. I have used the third and fourth editions mainly.

17. [Emerson], "A Man of Courage," *Treatise*, 205. From internal evidence I conclude that Emerson wrote this piece of self-praise and had it published. His correspondence reveals many instances in which he paid to have essays published in various journals. See James Emerson Collection, Merrimack Valley Textile Museum, North Andover, Massachusetts.

18. Emerson, "Engineers," *Treatise*, 4th ed., 40.

19. Emerson, "The Emerson Weir Tables," Ibid., 383–444.

20. Ibid., 383, 418.

21. "Wheel Testing," *Emerson's Turbine Reporter*, n.s., 1 (October 1874): 2.

22. Ibid.

23. Ibid.

24. Emerson, "Engineers," *Treatise*, 4th ed., 39.

25. A. M. Swain to James Emerson, *Emerson's Turbine Reporter*, n.s., 3 (no. 3, 1878): 4.

26. T. H. Risdon to the Editor, *Emerson's Turbine Reporter*, n.s., 2 (October 1875): 4.

27. Emerson to McCormick, February 27, 1892, and McCormick to Emerson, March 1, 1892, in Emerson, *Treatise*, 4th ed., 56. McCormick did in fact make use of the vernacular scientific tradition in his work; his statement can be read as true or false depending upon the interpretation given the phrase "old textbooks."

28. There is a brief summary of McCormick's work (and that of Swain) in John R. Freeman, "General Review of Current Practice in Water Power Production in America," *Transactions of the First World Power Conference*, 5 vols. (London: n.p., n.d.), vol. 2: 379, 380. Though there is no doubt about McCormick's effective use of science, his major creative contributions could not have been taken from textbooks; they involved the use by McCormick of scientific methods to address novel and important problems in design.

29. Emerson's notebooks recording his tests show that in many cases turbines that ended with very high efficiencies achieved these results only after many months of tinkering and testing one modification after another. See the three notebooks recording turbine tests, "Lowell Flume Tests, 1869–1874," "Holyoke Test Flume—1874–1879," and "Holyoke Test Flume, 1879–1880," James Emerson Collection, Merrimack Valley Textile Museum, North Andover, Massachusetts.

30. Emerson, "The Testing System," *Treatise*, 4th ed., 34.

31. Emerson, *Treatise*, 4th ed., 7.

32. *Emerson's Turbine Reporter* 2, no. 2 (October 1875): 3; no. 3 (January 1876): 3; no. 4 (n.d., 1876): 7.

33. "The Evolution of One of Emerson's Patents" and "Ship's Windlass," in Emerson, *Treatise*, 4th ed., 160–61. Though inferior to wrought iron and steel in tension, cast iron, if sufficiently massive, could bear considerable tensile force, as shown by the use of cast iron beams in British and in later American textile mills.

34. "The Evolution of One of Emerson's Patents," Ibid., 160. Emerson's device combined well-known elements: he combined the traditional ship's windlass with geared hoisting machinery and made both of cast iron.

35. "The American Ship Windlass Company," Ibid., 167.
36. "The Evolution of One of Emerson's Patents," and "Ship's Windlass," Ibid., 160–67.
37. Ibid., 150–59.
38. Hindle, *Emulation and Invention*, 12–14. Of course scientific technology does not exclude common sense and practical knowledge. To Emerson, however, practical common sense and craft experience were enough in themselves, without any aid from science.
39. Ibid., 273–81.
40. Ibid., 161.
41. Ibid., 131–32. The invention consisted of two nested stools combined into one, so that a second stool could be provided enabling two players to play the same piano. Both technically and socially the mustache cup and the duplex piano stool appear comparable. In both cases Emerson combined well-known means to meet a rather marginal social need.
42. T. W. Graham, "Percussion and Reaction," *Milling* 2 (January 1893): 149.
43. William Kennedy, "Direct Action and Reaction," *American Miller* 40 (Feb. 1, 1883): 60.
44. Graham, "Percussion and Reaction," 150.
45. C. R. Tompkins, "Water Wheels A Question of Efficiency," *Milling* 1 (October 1892): 496–97.
46. C. R. Tompkins, "Water Wheel Discussions," *Milling* 2 (May 1893): 501.
47. C. R. Tompkins, "Percussion and Reaction Wheels," *Milling* 2 (February 1893): 233.
48. T. W. Graham, "The Discussion of the Turbine," *Milling* 2 (March 1893): 364.
49. For examples of his many attacks on Christianity, see "The Day of the Church," "Superstition, Idolatry, or Worship, Which?" and the series of essays and cartoons starting with the title, "Myths of Prehistoric Times of Unknown and Unknowable Origin," Ibid., 122, 126, 127, 175, 364–79; for examples of attacks on marriage, see "Marriage, Divorce, Nudity," Ibid., 140–41.
50. "The Christian God" (cartoon), "Christianity," and "What Good Has Christianity Ever Done?" Ibid., 376–79.
51. "The Day of the Man," Ibid., 123.
52. Emerson, "Expurgation and Pretension," Ibid., 31.
53. Ibid., 33.
54. Emerson, "Progress in Medicine," Ibid., 125.
55. "The Law," Ibid., 26.
56. Emerson, "Rotten Statutes," Ibid., 31.
57. Emerson, "The Law Antagonistic to Knowledge and Justice," Ibid., 17
58. "Railroad Suggestions," Ibid., 86.
59. "The Republican Party," Ibid., 172
60. "The Protective Tariff," Ibid., 170.
61. "American Turbine," Ibid., 226–27.
62. Rudolph Camerer, *Vorlesungen Ueber Wasserkraftnmaschinen* (Leipzig: n.p., 1914): 295.

6. Race-ism and the City

1. Turner's essay, published originally in 1893, may be conveniently found in Frederick Jackson Turner, *The Frontier in American History* (New York: Macmil-

lan, 1920), 1–38. For Turner and the new emphasis on place in social theory, however, see Henry D. Shapiro, "The Place of Culture and the Problem of Identity," in Allen Bateau, ed., *Appalachia and America: Autonomy and Regional Dependence* (Lexington: University Press of Kentucky, 1983), esp. 109–28. This Turner should not be confused with the second Turner, the one who wrote about the significance of sections and the Balkanization of America, a step in the process of embedding culture in place rather than race. See Shapiro, "Place of Culture," in his *Appalachia on Our Mind: Southern Mountains and Mountaineers in the American Consciousness* (Chapel Hill: University of North Carolina Press, 1978), 128–34, and Zane L. Miller, "Pluralizing America: Walter Prescott Webb, the Chicago School of Sociology, and Cultural Regionalism," in Robert B. Fairbanks and Kathleen Underwood, eds., *Essays on Sunbelt Cities and Recent Urban America* (College Station: Texas A&M University Press, 1990), 151–76. I have borrowed the term "race-ist" from Shapiro. Shapiro explains that social theory before the late nineteenth century made no essential connection between culture and place and people and place, though by the mid-nineteenth century it had become common to use place (merely) as a means of classifying a people who possessed a particular culture. Place, that is, was not yet identified as a factor that might influence a culture. Shapiro also identifies the mid-nineteenth century as the time in which the concept of culture came to mean a way of life, the possession of primitive and untutored individuals and societies, not merely, as before, the possession of cultivated and "mature" individuals and societies. See Shapiro, "The Place of Culture," 111–19.

2. For an analysis of race-ism and the city during the 1880s, see Zane L. Miller, "The Rise of the City," *Hayes Historical Journal* (Spring and Fall, 1980): 73–84, esp. 77–80. For a comparison of Turner and Du Bois that emphasizes their differences see William Toll, "W. E. B. Du Bois and Frederick Jackson Turner: The Unveiling and Preemption of America's 'Inner' History," *Pacific Northwest Quarterly* 65 (April 1974): 66–78.

3. Du Bois in 1901 also published a series of articles (not discussed in this essay) that compared the Negro problem in New York, Philadelphia, and Boston. These articles have been collected and published as W. E. B. Du Bois, *The Black North in 1901: A Social Study* (New York: Arno Press, 1969).

4. W. E. B. Du Bois, *The Philadelphia Negro: A Social Study* (1899; reprint, New York: Schocken, 1967).

5. See Du Bois, *The Philadelphia Negro,* 73–82, especially 76–79, and Du Bois' study of the Negro race in the "metropolis" of Prince Edward County, Virginia, "The Negroes of Farmville, Virginia; A Social Study," United States Department of Labor Bulletin No. 14 (January 1899), 1–5.

6. Du Bois, *The Philadelphia Negro,* 176–77, 316–19.

7. W. E. B. Du Bois, *The Conservation of Races* (1897; reprint, New York: Arno Press, 1969), 5–15.

8. W. E. B. Du Bois, "Strivings of the Negro People," *Atlantic Monthly* 80 (July 1987): 194–98; *The Philadelphia Negro,* 73–74.

9. Du Bois, *The Conservation of Races.*

10. Du Bois, "Strivings."

11. For the proposal, see W. E. B. Du Bois, "The Study of the Negro Problems," *Annals of the American Academy of Political and Social Science* 11 (January–June 1898): 720, and his "Conditions of the Negro in Various Cities," United States

Department of Labor Bulletin No. 10 (May 1897), 257–369. On the late nine-teenth-century tendency to regard essences as effectively hidden from view, as practically unknowable through direct observation, and the consequences of the emergence of that view for the use of specimens and data in the social sciences, see Henry D. Shapiro's review of Michael M. Sokal, *An Education in Psychology: James Mckeen Cattell's Journals and Letters from Germany and England, 1880–1888* (Cambridge, Mass.: MIT Press, 1981), in *Technology and Culture* 25 (April 1984): 374–76, esp. 376.

12. Du Bois, *Conservation of Races,* 8. On the community of purpose in turn of the century social theory as the third stage of civilization after the community of blood and the community of territorial area, see Shapiro, "The Place of Cul-ture," 121, 135 (n8).

13. In addition to the works of Shapiro and Miller cited above, see (on the shift from race-based to place-based conceptions of culture) Andrea Tuttle Kornbluh, "From Culture to Cuisine: Twentieth-Century Views of Race and Ethnicity in the City," in Howard Gilette, Jr. and Zane L. Miller, eds., *American Urbanism: A Historiographical Review* (Westport, Conn.: Greenwood, 1987), 49–58, and her "The Bowl of Promise: Social Welfare Planners, Cultural Pluralism and the Metropolitan Community, 1911–1953" (Ph.D. diss., University of Cincinnati, 1988).

14. This is the view of one of the biographers of Du Bois, who claimed that the race-ism of Du Bois became the mainstream of his thought after the onset of the depression of the 1930s. See Francis L. Broderick, *W. E. B. Du Bois: Negro Leader in a Time of Crisis* (Stanford: Stanford University Press, 1959): 168–69.

15. This is the view of another biographer of Du Bois, who cites a 1958 publication for the definition of cultural pluralism. This biographer did not discuss or men-tion *The Conservation of Races.* See Elliott M. Rudwick, *W. E. B. Du Bois: Propa-gandist of the Negro Protest* (1960; reprint, New York: Arno Press, 1968), 36, 320 (n45).

16. This is the view of E. Digby Baltzell, expressed on pp. xxv and xxvi of his introduction to the 1967 edition of *The Philadelphia Negro,* and of Drake and Cayton. See St. Clair Drake and Horace A. Cayton, *Black Metropolis: A Study of Negro Life in a Northern City* (1945; reprint, New York: Harcourt, Brace & World, 1970), 787–88.

17. Joseph P. DeMarco in *The Social Thought of W. E. B. Du Bois* (Lanham, Md.: University Press of America, 1983) also describes the social theory of the young Du Bois as race-ist. DeMarco did not, however, locate the young Du Bois as one among many social theorists giving a greater emphasis on place while preserv-ing the primacy of race, identify Du Bois' view of the stages of civilization (as opposed to the stages in the development of the genius of a race,) discuss the appearance of the "medieval" concept in *The Philadelphia Negro,* or note the displacement during the 1920s of the idea of race-based culture by the idea of place-based culture.

18. For a biography that depicts the young (and the older) Du Bois as a radical democratic opponent of American capitalism and a cultural pluralist whose so-cial theory contained no biological content, see Manning Marable, *W. E. B. Du Bois: Black Radical Democrat* (Boston: Twayne, 1986), ix, 25–26, 35–38. Marable, like Broderick, Rudwick and Baltzell, also overlooks the new emphasis on place in turn of the century social theory, Du Bois' view of the stages of civilization, the "medieval" concept in *The Philadelphia Negro,* and the displacement during

the 1920s of the idea of race-based culture by the idea of place-based culture. An argument similar to Marable's may be found in Dan S. Green and Earl Smith, "W. E. B. Du Bois and the Concepts of Race and Class," *Phylon* 41, no. 5 (December 1983): 262–72.

7. The German-American Science of Racial Nutrition

1. Ellen H. Richards to Mary Hinman Abel, May 26, 1893, Ellen H. Richards Papers, American Home Economics Association, Washington, D.C.
2. Biographical information on Richards is sparse; perhaps the following will be useful: Caroline L. Hunt, *The Life of Ellen H. Richards* (Boston: Whitcomb and Barrows, 1918); Robert Clarke, *Ellen Swallow: The Woman Who Founded Ecology* (Chicago: Follett Publishing Company, 1973). Dolores Hayden, *The Grand Domestic Revolution: A History of Feminist Designs for American Homes, Neighborhoods, and Cities* (Cambridge: MIT Press, 1981), 150–79, discusses Richards's work in nutrition. The standard biography of Atkinson is Harold Francis Williamson, *Edward Atkinson: The Biography of an American Liberal, 1827–1905* (Boston: Old Corner Book Store, 1934).
3. Mary Hinman Abel, *Practical Sanitary and Economic Cooking Adapted to Persons of Moderate and Small Means. The Lomb Prize Essay* (Rochester, N.Y.: American Public Health Association, 1890).
4. A useful book on this entire problem is Harvey A. Levenstein, *Revolution at the Table: The Transformation of the American Diet* (New York: Oxford University Press, 1988), 44–59, 72–85, and passim. See also Levenstein, "The New England Kitchen and the Origins of Modern American Eating Habits," *The American Quarterly* 32 (Fall 1980): 369–86.
5. Edward Atkinson, "The Food Question in America and Europe," *Century Illustrated Magazine* 33 (1886–87): 238–48 (hereafter cited as *Century*); Atkinson, "The Relative Strength and Weakness of Nations," *Century*, 423–35, 613–21; Atkinson, "The Margin of Profits," *Century*, 923–31. See also Williamson, *Edward Atkinson*, 230–35.
6. Levenstein, "The New England Kitchen," 369–86; Levenstein, *Revolution at the Table*, 44–59.
7. See Alan I Marcus, *Agricultural Science and the Quest for Legitimacy: Farmers, Agricultural Colleges and Experiment Stations, 1870–1890* (Ames: Iowa State University Press, 1985), and Marcus, "Setting the Standard: Fertilizers, State Chemists and Early National Commercial Regulation," *Agricultural History* 61 (1987): 47–73.
8. Carl von Voit, *Physiologie Des Allgemeinen Stoffwechsels und Der Ernaehung, in Handbuch Der Physiologie,* Herausgegeben von Dr. L. Hermann (Leipzig: I. E. W. Vogel, 1881); Max Rubner, "Die Quelle der thierschen Waerme," *Zeitschrift Biologie* 30 (1894): 73–86; Rubner, *Die Gesetze des Energieverbruchs bie der Ernaehung* (Leipzig: Franz Deuticke, 1902), trans. Allan Markoff and Alex Sandri-White as *The Laws of Energy Consumption in Nutrition* (Bethesda, Md.: United States Army Research Institute of Environmental Medicine, 1969). I used the translation.
9. Wilbur O. Atwater, "On the Acquisition of Atmospheric Nitrogen by Plants," *American Chemical Journal* 6 (1885): 365–88, expresses his views on nitrogen and plants.
10. Wilbur O. Atwater, "The Chemistry of Foods and Nutrition," 6 parts: pt. 1,

Century 34 (1887–88): 54–74; pt. 2, "How Food Nourishes the Body," Ibid., 237–51; pt. 3, "The Potential Energy of Food," Ibid., 397–405; pt. 4, "The Digestibility of Food," Ibid., 733–40; pt. 5, "The Pecuniary Economy of Food," *Century* 35 (1888): 437–46; pt. 6, "Foods and Beverages," *Century* 36 (1888): 257–64.

11. Atwater's attitudes on this conflict were represented in Wilbur O. Atwater, *The What and Why of Agricultural Experiment Stations,* U.S. Office of Experiment Stations, Farmers Bulletin no. 16 (Washington, D.C.: Government Printing Office, 1889); Alan I Marcus, *Agricultural Science and the Quest for Legitimacy,* has an excellent discussion of this controversy. The standard biography of Morton is James C. Olson, *J. Sterling Morton* (Lincoln: University of Nebraska Press, 1942); see pp. 349–62 for a discussion of Morton as Secretary of Agriculture. The delectable "farm the Farmer" quote is in Horace Samuel Merrill, *Grover Cleveland and the Democratic Party* (Boston: Little, Brown, and Company, 1957), 170. Neither of the "standard" biographies of Cleveland has much of use about Morton as secretary of agriculture: Robert McElroy, *Grover Cleveland, the Man and the Statesman: An Authorized Biography,* vol. 2 (New York: Harper and Brothers, 1923), 6, 77, and Allan Nevins, *Grover Cleveland: A Study in Courage* (New York: Dodd and Mead, 1932) 514, 521.

12. Edward Atkinson, *Suggestions for the Establishment of Food Laboratories in Connection with the Agricultural Experiment Stations of the United States,* U.S. Office of Experiment Stations, Bulletin no. 17 (Washington, D.C.: Government Printing Office, 1893), quote at p. 9.

13. Ibid., p. 19.

14. Wilbur O. Atwater and F. G. Benedict, *A Respiration Calorimeter with Applicancies for the Direct Determination of Oxygen* (Washington, D.C.: Carnegie Institution of Washington, 1905), describes the apparatus and its use.

15. *U.S. Statutes at Large* 28 (1895): 271.

16. On Richards's later career in home economics, see Hunt, *The Life of Ellen H. Richards,* 259–300; Clarke, *Ellen Swallow,* 167–78; Emma Seifrit Weigley, "It Might Have Been Euthenics: The Lake Placid Conferences and the Home Economics Movement," *The American Quarterly* 26 (1974): 79–96; Mary Hinman Abel, et al., "The Home Economics Movement in the United States," *The Journal of Home Economics* 3 (1911): 323–417. For Abel's continuation of Richards's propagandizing in home economics after Richards's death, see, for example, Mary Hinman Abel, *Successful Family Life on the Moderate Income* (Philadelphia: Lippincott, 1921), passim.

17. The Adams Act *U.S. Statutes at Large* 34 (1906): 63. See also ibid. (1906): 694 and ibid. (1907): 2907, for the wrap-up of the human nutrition studies and their political relations to the Adams Act.

18. Wilbur O. Atwater and C. F. Langworthy, *A Digest of Metabolism Experiments in Which the Income and Outgo Was Determined,* U.S. Office of Experiment Stations, Bulletin no. 45 (Washington, D.C.: Government Printing Office, 1897).

19. Wilbur O. Atwater and E. B. Rosa, *Description of a New Respiration Calorimeter and Experiments on the Conservation of Energy in the Human Body,* U.S. Office of Experiment Stations, Bulletin no. 63 (Washington, D.C.: Government Printing Office, 1899), 89. Rubner's work was specifically cited here and elsewhere in Atwater's writings on the subject.

20. Wilbur O. Atwater, *Methods and Results of Investigations on the Chemistry and*

Economy of Food, U.S. Office of Experiment Stations, Bulletin no. 21 (Washington, D.C.: Government Printing Office, 1895), quote at p. 9.

21. Wilbur O. Atwater, Charles D. Woods, and F. G. Benedict, *Report of Preliminary Investigations on the Metabolism of Nitrogen and Carbon in the Human Organism with a Respiration Calorimeter of Special Construction,* U.S. Office of Experiment Stations, Bulletin no. 44 (Washington, D.C.: Government Printing Office, 1897); Atwater and A. P. Bryant, *The Chemical Composition of American Food Materials,* U.S. Office of Experiment Stations, Bulletin no. 28. (Washington, D.C.: Government Printing Office, 1899)

22. Whitman H. Jordan, *Dietary Studies at the Maine State College in 1895,* U.S. Office of Experiment Stations, Bulletin no. 37 (Washington, D.C.: Government Printing Office, 1897).

23. Edward B. Voorhees, *Food and Nutrition Investigations in New Jersey in 1895 and 1896,* U.S. Office of Experiment Stations, Bulletin no. 35 (Washington, D.C.: Government Printing Office, 1896).

24. H. B. Gibson, S. Calvert, and D. W. May, *Dietary Studies at the University of Missouri in 1895,* U.S. Office of Experiment Stations, Bulletin no. 31 (Washington, D.C.: Government Printing Office, 1896). Isabel Bevier also suggested the correlation between income and quality of nutrition in *Nutritional Investigations in Pittsburgh, Pa., 1894–1896,* U.S. Office of Experiment Stations, Bulletin no. 52 (Washington, D.C.: Government Printing Office, 1898).

25. Charles E. Wait, *Dietary Studies at the University of Tennessee in 1895, with comments by W. O. Atwater and Charles D. Woods,* U.S. Office of Experiment Stations, Bulletin no. 29 (Washington, D.C.: Government Printing Office, 1896), quote at p. 44.

26. Wilbur O. Atwater and Charles D. Woods, *Dietary Studies in New York City in 1895 and 1896,* U.S. Office of Experiment Stations, Bulletin no. 46 (Washington, D.C.: Government Printing Office, 1898), quotes at pp. 10, 33.

27. Ibid., quotes at pp. 14, 65.

28. Wilbur O. Atwater and A. P. Bryant, *Dietary Studies in Chicago in 1895 and 1896, Conducted with the Cooperation of Jane Addams and Caroline L. Hunt, of Hull House,* U.S. Office of Experiment Stations, Bulletin no. 55 (Washington, D.C.: Government Printing Office, 1898).

29. Arthur Goss, *Dietary Studies in New Mexico in 1895,* U.S. Office of Experiment Stations, Bulletin no. 40 (Washington, D.C.: Government Printing Office, 1897); Goss, *Nutritional Investigations in New Mexico in 1897,* U.S. Office of Experiment Stations, Bulletin no. 54 (Washington, D.C.: Government Printing Office, 1898).

30. Wilbur O. Atwater and Charles D. Woods, *Dietary Studies with Reference to the Food of the Negro in Alabama,* U.S. Office of Experiment Stations, Bulletin no. 38 (Washington, D.C.: Government Printing Office, 1897), 7.

31. Hamilton Cravens, "Establishing the Science of Nutrition at the United States Department of Agriculture: Ellen Swallow Richards and Her Allies," *Agricultural History* 64 (Spring 1990): 122–33.

32. Levenstein, *Revolution at the Table,* 44–60, 72–108, provides a larger context for these events, but his understanding of the history of science, here and elsewhere, is less than ideal.

33. E. V. McCollum, *The Newer Knowledge of Nutrition: The Use of Foods for the Preservation of Vitality and Health* (New York: Macmillan Co., 1918), v–viii, 1–34, and passim.

34. Subsequent editions of McCollum's *Newer Knowledge,* as in 1922, 1925, and later, make this abundantly clear. Unfortunately we lack a good, solid, history of nutritional science commencing with the late nineteenth century, save for McCollum's historical statements in his writings, which may be considered as having at least a trace element of special pleading in them.
35. Hamilton Cravens, *The Triumph of Evolution: The Heredity-Environment Controversy, 1900–1941* (Baltimore: Johns Hopkins University Press, 1988), discusses disciplinary competition in science in that era. Alan I Marcus and Howard P. Segal, *Technology in America: A Brief History* (San Diego: Harcourt Brace Jovanovich, 1989) cover changes in world views and in agricultural science in the 1870–1920 and 1920–50 eras.

8. The Case of the Manufactured Morons

1. Marie Skodak, *Children in Foster Homes: A Study of Mental Development,* University of Iowa Studies, Studies in Child Development, vol. 16, no. 1 (Iowa City: The University, 1939). This article is adapted from my *Before Head Start: The Iowa Station and America's Children* (Chapel Hill: University of North Carolina Press, 1993).
2. See Hamilton Cravens, *The Triumph of Evolution: The Heredity-Environment Controversy, 1900–1941* (Baltimore: Johns Hopkins University Press, 1988), 224–65.
3. Harold M. Skeels, Ruth Updegraff, Beth L. Wellman, and Harold M. Williams, *A Study of Environmental Stimulation: An Orphanage Preschool Project,* University of Iowa Studies, Studies in Child Welfare, vol. 15, no. 4 (Iowa City: The University, 1938).
4. See, for example: Beth L. Wellman, "Mental Growth from Preschool to College," *Journal of Experimental Education* 6 (1937): 127–38.
5. George D. Stoddard, "Some New Light on Human Intelligence," *California Journal of Secondary Education* 14 (1939): 490–94.
6. On the early history of child development, see, for example: Robert R. Sears, *Your Ancients Revisited: A History of Child Development* (Chicago: University of Chicago Press, 1975), passim. See also Hamilton Cravens, "Child-Saving in the Age of Professionalism, 1915–1930," in Joseph M. Hawes and N. Ray Hiner, eds., *American Childhood* (Westport, Conn.: Greenwood Press, 1985): 415–88. Much of what follows in this article is based on my research on the history of the field, most particularly my *Before Head Start: The Iowa Station and America's Children* (Chapel Hill: University of North Carolina Press, 1993).
7. Florence L. Goodenough and John E. Anderson, *Experimental Child Study* (New York: D. Appleton and Co., 1931), passim. I am especially indebted to Professor Albert E. Siegel and Dr. Bernadine Barr of Stanford University for many stimulating and enjoyable discussions about the intellectual history of developmental science. See also Hamilton Cravens, "Behaviorism Revisited: Developmental Science, the Maturation Theory, and the Biological Basis of the Human Mind, 1920s–1950s," in Keith R. Benson, Jane Maienschein, and Ronald Rainger, eds., *The Expansion of American Biology* (New Brunswick: Rutgers University Press, 1991), 133–63. In *The Triumph of Evolution* I have argued that this was precisely the result of the heredity-environment controversy although I gather many readers assume (and wrongly so) that the controversy's resolution led to the "triumph of culture."

8. Goodenough and Anderson, *Experimental Child Study,* 505–15.
9. Alice Leahy, *Nature-Nurture and Intelligence,* Genetic Psychology Monographs, vol. 17, no. 4 (1935): 235–309. On the importance of the group, and on the meaning of group membership for the individual, see, for example, Hamilton Cravens, "History of the Social Sciences," in Sally G. Kohlstedt and Margaret W. Rossiter, eds., *Historical Writing on American Science: Perspectives and Prospects* (Baltimore: Johns Hopkins University Press, 1986), 183–207.
10. George D. Stoddard, "Extending the Schools Downward," *Educational Administration and Supervision* 15 (1929): 581–92; Stoddard, "Some Current Ideas in Nursery Education," *School and Society* 35 (1932): 277–80; Stoddard, "A Survey of Nursery School Costs," *Journal of Educational Research* 26 (1932–33): 354–59; Stoddard, "What Shall Be the Function of the National Association for Nursery Education?" *Proceedings of the Fifth Annual Conference of the National Association for Nursery Education,* Toronto, Ontario, October 26–28, 1933 (Boston, 1933), 6–7.
11. Beth L. Wellman, "Some New Bases for Interpretations of the IQ," *Pedagogical Seminary and Journal of Genetic Psychology* 41 (1932): 116–26; Wellman, "The Effect of Preschool Attendance upon the IQ," *Journal of Experimental Education* 1 (1932–33): 48–69. See also the *Cedar Rapids Gazette,* December 31, 1931.
12. Beth L. Wellman, "Growth in Intelligence Under Differing School Environments," *Journal of Experimental Education* 3 (1934–35): 59–83.
13. Hubert S. Coffey and Beth L. Wellman, "The Role of Cultural Status in Intelligence Changes of Preschool Children," *Journal of Experimental Education* 5 (1936–37): 191–202, quote at p. 202.
14. See, for example, Beth L. Wellman, "Mental Growth from Preschool to College," *Journal of Experimental Education* (1937–38): 127–38; Wellman, "Guiding Mental Development," *Childhood Education* 15 (1938–39): 108–12; Wellman, "Our Changing Concept of Intelligence," *Journal of Consulting Psychology* 2 (1938): 97–107.
15. I am indebted to Dr. Marie Skodak Crissey for discussing certain aspects of the orphanage studies with me at the 1983 meeting of the Society for Research in Child Development, Detroit, Michigan. See George D. Stoddard, "Notes on Reorganization and Support of Station Activities 1934–35 and 1935–36," April 5, 1934, six typescript pages, p. 6, Papers of the Presidents of the University of Iowa, University of Iowa Archives, Iowa City, Iowa; Harold M. Skeels, "Mental Development of Children in Foster Homes," *Pedagogical Seminary and Journal of Genetic Psychology* 49 (1936): 91–106; Skeels, "The Relation of the Foster Home Environment to the Mental Development of Children Placed in Infancy," *Child Development* 7 (1936) 1–5; *Des Moines Register,* June 18, 1936.
16. Skeels et al. *Study of Environmental Stimulation,* passim.
17. Harold M. Skeels and Harold Dye, "A Study of the Effects of Differential Stimulation on Mentally Retarded Children," *Proceedings of the American Association of Mental Deficiency* 44 (1939): 114–36; Skeels, "A Study of the Effects of Differential Stimulation on Mentally Retarded Children: A Follow-Up Report," *American Journal of Mental Deficiency* 46 (1942): 340–50.
18. Henry L. Minton, *Lewis M. Terman: Pioneer in Psychological Testing* (New York: New York University Press, 1988), 191–201.
19. The yearbook in question is Guy M. Whipple, ed., *The Thirty-Ninth Yearbook of the National Society for the Study of Education. Intelligence: Its Nature and Nurture.*

Part I. Comparative and Critical Exposition. Part II. Original Studies and Experiments (Bloomington: Public School Publishing Co., 1940). The details of the controversy and fuller citations are in Minton, *Lewis M. Terman,* 191–201.

20. I am deeply indebted to Bernadine Barr of Stanford University for illuminating discussions of statistical method; see also Lee J. Cronbach, "The Two Disciplines of Scientific Psychology," *American Psychologist* 12, no. 11 (November 1958): 671–84.

21. Quinn McNemar, "A Critical Examination of the University of Iowa Studies of Environmental Influences upon the IQ," *Psychological Bulletin* 37 (1940): 63–92; McNemar, "More on the Iowa IQ Stories," *Journal of Psychology* 10 (1940): 237–40; Beth L. Wellman and E. L. Pegram, "Binet IQ Changes of Orphanage Preschool Children: A Re-Analysis," *Journal of Genetic Psychology* 65 (1944): 239–63; Beth L. Wellman, Harold M. Skeels, and Marie Skodak, "Review of McNemar's Critical Examination of Iowa Studies," *Psychological Bulletin* 37 (1940): 93–111. The final salvo was Quinn McNemar, "Note on Wellman's Re-Analysis of IQ Changes of Orphanage Preschool Children," *Journal of Genetic Psychology* 67 (1943): 215–19. To this day, psychologists still do not have a way to represent the individuality or individuation that the Iowans were discussing, save through essentially descriptive techniques, either noting if a large number of individuals do not approximate a group measurement or, going to the other extreme, constructing individual slope lines for each individual studied. This would appear to have fascinating cultural as well as technical implications worthy of further exploration and explication.

22. Florence L. Goodenough to Lewis M. Terman, February 22, 1940, folder 13, box 14, *Lewis M. Terman Papers,* Stanford University Archives, Stanford University Libraries, Stanford, California.

23. J. McVicker Hunt, *Intelligence and Experience* (New York: Ronald Press, 1961); Benjamin Bloom, *Stability and Change in Human Characteristics* (New York: Wiley and Sons, 1964). There is a fairly large literature on Head Start. A useful point of departure is Edward Zigler and Jeanette Valentine, eds., *Project Head Start: A Legacy of the War on Poverty* (New York: Free Press, 1979).

24. Harold M. Skeels, *Adult Status of Children with Contrasting Early Life Experiences: A Follow-Up Study,* Monographs of the Society for Research in Child Development (Chicago: University of Chicago Press, 1966) serial no. 105 (1966), vol. 31, no. 3, passim.

25. See Cravens, "History of the Social Sciences."

9. Responding to the Airplane

1. Eric Monkonnen, *America Becomes Urban: The Development of U.S. Cities and Towns, 1780–1980* (Berkeley and Los Angeles: University of California Press, 1988), 162.

2. U.S. Department of Interior, Bureau of Census, *Fifteenth Census of the United States, 1930: Population,* vol. 1, 1056; William Neil Black, "Empire of Consensus: City Planning, Zoning and Annexation in Dallas, 1900–1960," (Ph.D. diss., Columbia University, 1982), 286.

3. Stanley H. Scott and Levi H. Davis, *A Giant in Texas: History of the Dallas–Fort Worth Regional Airport Controversy, 1911–1974* (Quanah, Tex.: Nortex Press, 1974), 3–4.

4. *Dallas Morning News (DMN),* July 7, 1927, November 30, 1927.

5. *DMN,* December 15, 1927, July 8, 1927 (quote).

6. *DMN,* June 30, 1928; Dallas, Texas, *Progress An Official Report of Municipal Achievement in Dallas,* 1934, 22.

7. Dallas, Texas Aviation Department, General Information, Dallas Love Field–Redbird Airport, Dallas Public Library (DPL); Scott and Davis, *Giant,* 3; American Public Works Association, *History of Public Works in the United States, 1776–1976* (Chicago: American Public Works Association, 1976), 192; *Dallas Dispatch-Journal,* November 21(?), 1939, Love Field Clipping File, DPL.

8. Donald R. Whitnah, *Safer Skyways: Federal Control of Aviation, 1926–1966* (Ames: Iowa State University Press, 1966), 166–67.

9. Jodi Goodwin, "Conflict and Compromise: The Regional Airport Experience in Dallas and Fort Worth," in Robert Ray Mason, Jr., and Craig W. Barnes, eds., *Essays in History: The E. C. Barksdale Student Lectures, 1987–1988* (Arlington, Tex.: Department of History, University of Texas at Austin, 1988), 246–47; *Fort Worth Star Telegram (FWST),* October 2, 1940; Scott and Davis, *Giant,* 5–6.

10. *FWST,* October 29, 1940.

11. Ibid., August 19, 1962; Scott and Davis, *Giant,* 6.

12. Scott and Davis, *Giant,* 6–7; *FWST,* September 10 and 16, 1941, October 17, 1941, January 8, 1942.

13. Robert Weer, "A Dream Dies," *Dallas Times Herald (DTH),* March 17, 1974; Scott and Davis, *Giant,* 7–8; *DMN,* March 1 and 2 (quote), 1943; *FWST,* March 23, 1943.

14. Roderick D. McKenzie, *The Metropolitan Community* (New York: McGraw-Hill, 1933), 69.

15. Zane L. Miller, "Pluralizing America: Walter Prescott Webb, Chicago School Sociology, and Cultural Regionalism," in Robert B. Fairbanks and Kathleen Underwood, eds., *Essays on Sunbelt Cities and Recent Urban America* (College Station: Texas A&M University Press, 1990), 162–63. Roderick McKenzie made a similar observation in 1933 when he wrote, "The supercommunity [metropolitan region] therefore absorbs varying members of separate local communities into its economic and cultural organizations. In this pattern, a dominant city that is dominant relative to secondary settlement functions as the integrating unit" (McKenzie, *Metropolitan Community,* 312).

16. Harland Bartholomew and Associates, *A Master Plan for Dallas, Texas,* Reports no. 1, 2, 5, and 10 (St. Louis, 1943); *DMN,* August 31, 1943.

17. Bartholomew and Associates, *Master Plan,* Report no. 13, p. 9.

18. Ibid., Report no. 5.

19. *DTH,* February 1, 1944.

20. Bartholomew and Associates, *Master Plan,* Report no. 7.

21. Ibid.; Dallas, Texas, "Report on Expansion Facilities at Love Field Airport," (May 1944), 3–5, Aviation Dept., Love Field. E. O. Pearson of Harland Bartholomew and Associates directed the report, and the Dallas engineering firm of Rollins and Forrest drew it up.

22. Bartholomew and Associates, Report no. 6, "Transportation Facilities," 43; Report no. 13, "Administrative Policy and Practice" (March 1945), 9.

23. Scott and Davis, *Giant,* 11; *DMN,* November 5, 1943, December 5, 1943, April 29, 1944.

24. Scott and Davis, *Giant,* 11–12; *DTH,* November 18, 1947.

25. *History of Public Works,* 196–97.
26. Scott and Davis, *Giant,* 18; *DMN,* October 30, 1947.
27. *DMN,* 1948, February 3, March 6, June 1 and 2, July 4 and 7, September 16; *DTH,* February 11, 1948.
28. *DTH,* April 15, 1951, June 26, 1951, August 5, 1951. Although Carpenter's support of the airport may be solely a result of his progressive vision, it should be noted that his family owned much real estate near the airport and would undoubtedly benefit from Carter Field's development.
29. *DMN,* December 10, 1951, November 26, 1954, October 20, 1954; Andy DeShong, "The Dallas Chamber of Commerce: Its First Seventy Years" (typescript, *DPL*); interview with George Coker, September 12, 1989.
30. *DMN,* December 10, 1951; *New York Times,* April 1, 1949, March 13, 1969; James C. Buckley, Inc., "Report of the Future of Love Field Dallas, Texas with Respect to Service by Scheduled Air Carriers," March 1952 (typescript, *DPL*); Buckley, "The Effect of Airport Distance on Traffic Generation," *Journal of the Air Transport Division: Proceedings of the American Society of Civil Engineers* 82 (May 1956): 978-1–978-19.
31. Buckley, "Report of the Future," 10; *DMN,* November 27, 1954.
32. Buckley, "Report of the Future," 59–60, 10–11.
33. James C. Buckley, Inc., "Air Service Requirements for the City of Dallas, Texas. A Study for the Dallas Chamber of Commerce," October 15, 1952 (typescript, *DPL*), 47–55; DeShong, "Dallas Chamber of Commerce," 59–61.
34. Dallas, *City Council Minutes,* July 1, 1952, Resolution 522510; *DMN,* June 18, 1952, January 23 and 27, 1953.
35. *DMN,* January 26, 1953.
36. *DTH,* January 20, 1953, *DMN,* January 21, 24, 26, and 28, 1953.
37. James C. Buckley, Inc., "A Master Plan Study of Love Field," (New York, February 8, 1954), 1–2; *DTH,* December 28, 1954.
38. Scott and Davis, *Giant,* 23; *DMN,* November 22, 1954; "Aviation Rivals," *Aviation Week,* November 15, 1954, 29–30.
39. Draft of Proposed Report of Aviation Department to George Coker, appendix D; Dallas Chamber of Commerce, *Annual Report,* 9 (*DPL*); Scott and Davis, *Giant,* 28–29.
40. *DTH,* December 15, 1955, February 9, 1956; *DMN,* March 21, 1956, November 18, 1951; Scott and Davis, *Giant,* 34.
41. *DMN,* October 20, 1957; *DTH,* January 1958.
42. The first commercial jet landed at Love Field on November 10, 1958. Regular jet service started the following February (*DMN,* November 11, 1958). Dallas's longest runway at this time was 7,500 feet (Scott and Davis, *Giant,* 34).
43. *DMN,* March 31, 1959; *DTH,* November 13 and 22, 1959; Draft of Proposed Report of the Aviation Committee for George Coker, 1960.
44. *DTH,* October 9, 1962.
45. James C. Buckley, Inc., "Re-evaluation of the Master Plan for Love Field," (Dallas, Texas, 1960), 9.
46. Ibid., 87–88. It is important to remember that Buckley was not automatically opposed to regional airports; he had been the chief architect of regional airport development for the New York Authority in the late 1940s. The difference was that he saw that as a discrete region, whereas he did not view the Dallas-Fort Worth region as a discrete unit. For more on Buckley's activity for the Port of

New York Authority, see Jameson W. Doig, "Coalition-Building by a Regional Agency: Austin Tobin and the Port of New York Authority," in Clarence N. Stone and Heywood T. Sanders, eds., *The Politics of Urban Development* (Lawrence: Regents Press of Kansas, 1987), 73–104.

47. In March 1960, the FAA approved Fort Worth's request for $907,500 to lengthen Carter Field while at the same time denying a Dallas request. James Winchester, "The Great Dallas–Fort Worth Controversy," *Flying*, May 1961, 85; *DMN*, March 1, 1963.

48. Telegram to Najeeb E. Halaby from Earle Cabell, August 1962, Earle Cabell Papers, Southern Methodist University Library, Dallas, Texas; *DMN*, August 17, 1962; *DTH*, August 30, 1962.

49. Earle Cabell News Release, August 22, 1962; H. L. Nichols to membership of Chamber of Commerce, August 31, 1962; Love Field Advisory Committee Data, August 20, 1962, Cabell Papers.

50. *DMN*, July 7, 16, 17, and 18, 1963, September 20, 1963.

51. Ibid., July 31, 1963, August 5, 1963, September 18, 1963.

52. CAB, "Initial Decision of Ross I. Newman," April 7, 1964, Dockett 13959, 45, Cabell Papers; Scott and Davis, *Giant*, 49.

53. *DMN*, June 1, 1965; March 28, 1965.

54. Zane L. Miller, *Suburb: Neighborhood and Community in Forest Park, Ohio, 1935–1976* (Knoxville: University of Tennessee Press, 1981), 227.

55. David Bauer, "Metroplexed," *D, The Magazine of Dallas*, January 1975, 65; *FWST,* September 23, 1973; "Dallas/Fort Worth, the Southwest Metroplex: A New World Capitol," *Fortune*, October 1973, 59–62.

10. Unanticipated Aftertaste

1. Iowa State College Information Service, "New Feeding Research Means More Beef at Lower Cost for Consumer in Long Run," February 18, 1954; "Hormone Feeding Speeds up Gains in Beef Steers, Reduces Costs," February 18, 1954; and "Stilbestrol Experiments Promise Faster, Cheaper Beef Production," February 18, 1954 in Iowa State College Research Foundation Collection (hereafter cited as ISCRF), Burroughs General Correspondence from April 28, 1953 folder. Wise Burroughs, C. C. Culbertson, Edmund Cheng, W. H. Hale, and Paul Homeyer, "The Influence of Oral Administration of Diethylstilbestrol to Beef Cattle," *Journal of Animal Science* (hereafter cited as *J. Ani. Sci.*) 14 (1955): 1015–24; M. T. Clegg and F. D. Carroll, "Further Studies of the Anabolic Effect of Stilbestrol in Cattle as Indicated by Carcass Composition," *J. Ani. Sci.* 15 (1956): 37–43; Joe Kastelic, Paul Homeyer, and F. A. Kline, "The Influence of the Oral Administration of Diethylstilbestrol on Certain Carcass Characteristics of Beef Cattle," *J. Ani. Sci.* 15 (1956): 689–700; and Wise Burroughs, C. C. Culbertson, Joseph Kastelic, Edmund W. Cheng, and William H. Hale, "Oral Administration of Diethylstilbestrol for Growth and Fattening in Beef Cattle," *J. Ani. Sci.* 13 (1954): 978. The ISCRF is housed in the Special Collections Department, Parks Library, Iowa State University, Ames, Iowa.

2. Letter from Mrs. Robert Swenson, River Falls, Wisconsin, to Dr. Wise Burroughs, Iowa State College, February 19, 1954, ISCRF, Burroughs General Correspondence from April 28, 1953 folder.

3. Letter from Dr. Wise Burroughs, Iowa State College, to Mrs. Robert Swenson,

River Falls, Wisconsin, March 4, 1954, ISCRF, Burroughs General Correspondence from February 1, 1954 folder.
4. See, for instance, William F. Longgood, *The Poisons in Your Food* (New York: Simon and Schuster, 1960); James S. Turner, *The Chemical Feast* (New York: Grossman, 1970); Samuel S. Epstein, *The Politics of Cancer* (San Francisco: Sierra Club Books, 1978); "Panic Over Food Additives," *Washington Post,* August 10, 1975, p. B-6; "Who Regulates the Regulators?" *Washington Post,* June 24, 1976, p. E-1; "Does Beef-Fattener Leave Cancer Seeds in Meat?" *National Observer,* November 20, 1971, p. 10. This literature of advocacy is extensive. Also see, for example, Michael F. Jacobson, *Eater's Digest: The Consumer's Factbook of Food Additives* (Garden City: Doubleday, 1972); Ruth Winter, *Beware of the Food You Eat* (New York: Crown, 1971); Jacqueline Verrett and Jean Carper, *Eating May Be Hazardous to Your Health* (New York: Simon and Schuster, 1974); Beatrice Trum Hunter, *The Mirage of Safety: Food Additives and Federal Policy* (New York: Charles Scribners Sons, 1975); Harrison Wellford, *Sowing the Wind: A Report From Ralph Nader's Center for Study of Responsive Law on Food Safety and the Chemical Harvest* (New York: Grossman, 1972); and Environmental Defense Fund and Robert H. Boyle, *Malignant Neglect* (New York: Knopf, 1979). For similar statements issued somewhat earlier, see, for example, Franklin Bicknell, *Chemicals in Your Food and in Farm Produce: Their Harmful Effects* (New York: Emerson Books, 1961); Thomas Powell, "New Poisons Imperil Our Meat," *Organic Gardening and Farming* 6 (March 1959): 99–103; and "Harmful Drug in Meat?" *Consumer Research Magazine* 45 (May 1962): 29, 39. Only a handful of commentators actively supported chemical additives in food. See Melvin A. Benarde, *The Chemicals We Eat* (New York: American Heritage Press, 1971); and Elizabeth M. Whelan and Frederick J. Stare, *Panic in the Pantry: Food Facts, Fads and Fallacies* (New York: Atheneum, 1975). For FDA approval of DES as a cattle feed additive, see "Memo—Jack M. Curtis to J. Hauser, October 28, 1954," *U.S. Food and Drug Administration Papers,* housed in FDA Headquarters, Rockville, Md., accession no. 88–82–57, box 21, vol. 280; and "FDA Authorizes Sale of Stilbestrol Pre-Mix for Use in Cattle Feeds," *Feedstuffs* 26 (November 6, 1954): 1, 8.
5. "Suit Filed on Cattle Hormone," *Washington Post,* October 29, 1971, p. A-2; *Transcript of the Special Meeting on Diethylstilbestrol of the National Advisory Drug Committee of the Food and Drug Administration,* July 25, 1972, 110ff; Letter from Consumer Action Now, Inc. to Edward M. Kennedy, Chairman of the Senate Subcommittee on Health, in *Hearing Before the Subcommittee on Health of the Committee on Labor and Public Welfare, United States Senate,* July 20, 1972, 73–74; "Petition of the Drugs Out of Meat Committee to the Senate Subcommittee on Health," in *Hearing Before the Subcommittee on Health,* July 20, 1972, 98–99; Dorothy Cottrell, "The Price of Beef," *Environment* 13 (July–August 1971): 44–51; and Joan Arehart-Treichel, "Agonizing Over Food and Drugs: How Safe?" *Science News* 103 (June 2, 1973): 362–63.
6. For institutionalization of this perspective about science and scientists, see, for example, Alan I Marcus, "Setting the Standard: Fertilizers, State Chemists and Early National Commercial Regulation, 1880–1887," *Agricultural History* 61 (Winter 1987): 47–73, and "Professional Revolution and Reform in the Progressive Era: Cincinnati Physicians and the City Elections of 1897 and 1900," *Journal of Urban History* 5 (1979): 183–207; Hamilton Cravens, "Child-Saving in the

Age of Professionalism, 1915–1930," in Joseph M. Hawes and N. Ray Hiner, eds., *American Childhood: A Research Guide and Historical Handbook* (Westport, Conn.: Greenwood Press, 1985), 415–88; Edward H. Beardsley, *The Rise of the American Chemical Profession, 1850–1900* (Gainesville: University of Florida Press, 1964), passim; Daniel J. Kevles, *The Physicists* (New York: Vintage, 1979), 45–71 and passim; and Samuel P. Hays, *Conservation and the Gospel of Efficiency* (Cambridge: Harvard University Press, 1959), passim. For a broader examination of the idea of scientific absolutism, see Michel Foucault, *The Birth of the Clinic: An Archaeology of Medical Perception* (New York: Vintage, 1973), passim.

7. For representative critiques and sources, see, for example, "FDA in Cyclamate Muddle," *Science and Government Report* 4 (March 1, 1974): 6; "Public Interest Group Petitions for Ban of Nitrate and Nitrite in Meat," *Consumer News* 7 (December 1, 1977): 3; "Chemicals in Our Foods and Beverages," *Consumer Bulletin Annual* (1973): 4–7; G. O. Kermode, "Food Additives," *Scientific American* 226 (March 1972): 15–21; Thomas H. Jukes, "Antibiotics in Animal Feeds and Animal Production," *Bioscience* 22 (1972): 526–34; and *Hearings Before the Select Committee on Small Business of the United States Senate, on Food Additives: Competitive, Regulatory, and Safety Problems, January 13 and 14, 1977* (Washington, D.C.: Government Printing Office, 1977). There has been little scholarship on the history of DES as a cattle growth promoter. Best is Terry G. Summons, "Animal Feed Additives, 1940–1966," *Agricultural History* 42 (1968): 305–13; Orville Schell, *Modern Meat* (New York: Random House, 1984); and Joseph V. Rodericks, "FDA's Ban of the Use of DES in Meat Production: A Case Study," *Agriculture and Human Values* 3 (Winter–Spring 1986): 10–25. The environmental movement and war on cancer have been better served. See, for instance, Thomas R. Dunlap, *DDT: Scientists, Citizens, and Public Policy* (Princeton: Princeton University Press, 1981); Richard A. Rettig, *Cancer Crusade: The Study of the National Cancer Act of 1971* (Princeton: Princeton University Press, 1977); Nicholas H. Steneck, *The Microwave Debate* (Cambridge: MIT Press, 1984); James T. Patterson, *The Dread Disease: Cancer and Modern American Culture* (Cambridge: Harvard University Press, 1987); Samuel P. Hays, *Beauty, Health, and Permanence: Environmental Politics in the United States, 1955–1985* (New York: Cambridge University Press, 1987); and Benjamin W. Mintz, *OSHA: History, Law, and Policy* (Washington: BNA Books, 1984). Of interest is Peter Temin, *Taking Your Medicine: Drug Regulation in the United States* (Cambridge: Harvard University Press, 1980).

8. For announcement of the synthesis of DES, see Edward Charles Dodds, L. Goldberg, W. Lawson, and R. Robinson, "Oestrogenic Activity of Certain Synthetic Compounds," *Nature* 141 (1938): 247–48. For DES as a carcinogen, see Charles F. Geschickter, "Mammary Carcinoma in the Rat with Metastasis Induced by Estrogen," *Science* 89 (January 13, 1939): 35–37; and Michael B. Shimkin and Hugh Grady, "Mammary Carcinomas in Mice Following Oral Administration of Stilbestrol," *Proceedings of the Society for Experimental Biology and Medicine* 45 (October 1940): 246–48. For the early years of DES, see Susan Bell, "The Synthetic Compound Diethylstilbestrol (DES), 1938–1941: The Social Construction of a Medical Treatment" (Ph.D. diss., Brandeis University, 1980). For a hint of the relationship between Burroughs's work and estrogens in nature, see "Stilbestrol Hormone Tested with Lambs," *Iowa Farm Science* 9 (November 1954): 18. Also see C. P. Wilsie, H. D. Hughes, and J. M. Scholl,

"Ladino Clover For Rotation Pastures," *Iowa Farm Science* 6 (January 1952): 14–16; Edward W. Ruf, "In Vitro Study of Unidentified Factor(s) in Cattle Feeds Which Favorably Influence Digestion by Rumen Micro-organisms," (Ph.D. diss., Iowa State College, 1952); Eugene N. Francis, "Use of Stilbestrol in Fattening Rations," (master's thesis, Iowa State College, 1953); and Wise Burroughs, "Method of Treating Growing Beef Cattle and Sheep and Feed Materials for Use Therein, June 3, 1953," ISCRF, Burroughs Method of Treating Growing Beef Cattle folder. The last item is Burroughs's patent application.

9. Burroughs et al., "The Influence"; Martin Stob, F. N. Andrews, and M. X. Zarrow, "The Detection of Residual Hormone in the Meat of Animals Treated with Synthetic Estrogens," *American Journal of Veterinary Research* 15 (1954): 319–22; and Charles W. Turner, "Biological Assay of Beef Steer Carcasses for Estrogenic Activity Following the Feeding of Diethylstilbestrol at a Level of .10 mg. per Day in the Rations," *J. Ani. Sci.* 15 (1956): 13–24.

10. For the association, see, for example, Ernest J. Umberger, Daniel Banes, Frieda M. Kunze, and Silvia H. Colson, "Chemical Determination of Diethylstilbestrol Residues in the Tissues of Treated Chickens," *Journal of the Association of Official Analytical Chemists* (hereafter *JAOAC*) 46 (1963): 471–79; George Schwartzman, "Report on Drug Residues in Animal Tissues," *JAOAC* 51 (1968): 270–71; Anthony J. Malanoski, "Regulatory Control of Hormone Residues by Chemical Analysis," *JAOAC* 53 (1970): 226–28; David E. Coffin and Jean-Claude Pilon, "Gas Chromatographic Determination of Diethylstilbestrol Residues in Animal Tissues," *JAOAC* 56 (1973): 352–57; Carl Ponder, "Fluorometric Determination and Thin Layer Chromatographic Identification of Diethylstilbestrol in Beef Liver," *JAOAC* 57 (1974): 919–23; Edgar W. Day, Jr., Lynn E. Vanattam, and Robert F. Sleck, "The Confirmation of Diethylstilbestrol Residues in Beef Liver by Gas Chromatography-Mass Spectrometry," *JAOAC* 58 (1975): 520–24; Bernd Hoffman, "Use of Radioimmunoassay for Monitoring Hormonal Residues in Edible Animal Products," *JAOAC* 61 (1978): 1263. For the Delaney clause, see *Public Law* 85–929, 72 Statutes 1784, Sec. 409, Part C. For the DES amendment, see Public Law 87–781, Sec. 104, Part F. For grandfathered exceptions to the Delaney clause, see, for example, Jay Richter, "Washington Letter," *Nebraska Farmer* 102 (January 2, 1960): 10.

11. G. E. Mitchell, Jr., A. L. Neumann, and H. H. Draper, "Metabolism of Tritium-Labeled Diethylstilbestrol by Steers," *Journal of Agricultural and Food Chemistry* 7 (1959): 509–12; Ernest J. Umberger, Jack M. Curtis, and George H. Gass, "Failure to Detect Residual Estrogenic Activity in the Edible Tissues of Steers Fed Stilbestrol," *J. Ani. Sci.* 18 (1959): 221–26; "New Clearances for Stilbestrol," *Feed Age* 13 (March 1963): 88; "FDA Issues Requirements on Diethylstilbestrol," *Feed Age* 13 (April 1963): 26; and National Academy of Sciences/National Research Council, *The Use of Drugs in Animal Feeds: Proceedings of a Symposium* (Washington: National Academy of Sciences, 1969): 249, 266, 281–82.

12. "Discovery Leads to Tune-up DES," *Feedlot Management* 12 (January 1970): 58. This new material was 90 percent trans isomer and 10 percent cis isomer. Earlier DES had been roughly 70 percent trans isomer and 30 percent cis isomer.

13. "FDA Asked to OK Double Stilbestrol Limit," *Farm Journal* 93 (August 1969): B-13; "Tired of Consumerism," *Feedlot Management* 12 (September 1970): 11; Bob Dunaway, "New Level for Stilbestrol Is Approved," *Wallaces Farmer* 95 (Oc-

tober 10, 1970): 13; "Regulation of Food Additives and Medicated Animal Feeds," *Hearings Before a Subcommittee on Government Operations of the House of Representatives,* March 16, 17, 18, 29, and 30, 1971 (Washington, D.C.: Government Printing Office, 1971), 324–33, 371–82, 408–20; and George D. Lakata, "Report on Drug Residues in Animal Tissues," *JAOAC* 54 (1971): 281–82.

14. Arthur L. Herbst, Howard Ulfelder, and David C. Poskanzer, "Adenocarcinoma of the Vagina: Association of Maternal Stilbestrol Therapy with Tumor Appearance in Young Women," *New England Journal of Medicine* 284 (1971): 878–81. For a suggestive earlier paper, see Arthur L. Herbst and Robert F. Scully, "Adenocarcinoma of the Vagina in Adolescence," *Cancer* 25 (1970): 745–57. Most retrospective scholarship about DES has focused on the DES daughters controversy. See Roberta J. Apfel and Susan M. Fisher, *To Do No Harm: DES and the Dilemma of Modern Medicine* (New Haven: Yale University Press, 1984); Richard Gillam and Barton J. Bernstein, "Doing Harm: The DES Tragedy and Modern American Medicine," *Public Historian* 9 (Winter 1987): 57–82; Robert Myers, *DES: The Bitter Pill* (New York: Putnam, 1983); and Kenneth L. Noller and Charles R. Fish, "Diethylstilbestrol Usage: Its Interesting Past, Important Present, and Questionable Future," *Medical Clinics of North America* 58 (July 1974): 793–810.

15. "Drug Linked to Cancers in Daughters," *Washington Post,* October 27, 1971, pp. A-1 and A-6; "Cancer Chemical Is Found in Beef," *Washington Post,* October 9, 1971, p. A-2; *Federal Register* 36 (October 23, 1971): 20534; "Proxmire Bill Would Impose Ban on Cancer-Linked Cattle Hormone," *Washington Post,* November 9, 1971, p. A-3; "FDA Warning on DES," *Washington Post,* November 10, 1971, p. B-4; "U. N. Body Seeks Ban on Synthetic Hormone Used in Cattle Feed," *Washington Post,* November 11, 1971, p. F-6; and "Rules Seen Decreasing Chemical in Beef, Lamb," *Washington Post,* December 14, 1971, p. A-4. See also "Now It's Beef and Mutton," *Newsweek,* 78 (November 8, 1971): 85; and Kathryn S. Huss, "Maternal Diethylstilbestrol a Time Bomb for Child?" *Journal of the American Medical Association* 218 (December 6, 1971): 1564–65.

16. "M. Adrian Gross and R. L. Gillespie to Dr. Leo Friedman, Director, Division of Toxicology, Bureau of Foods, FDA and Dr. Daniel Banes, Director, Office of Pharmaceutical Research and Testing, Bureau of Drugs, FDA, December 5, 1971," "M. Adrian Gross, DVM, Bureau of Drugs, FDA, to J. Richard Crout, M. D., Deputy Director, Bureau of Drugs, FDA, January 10, 1972," and "Nathan Mantel, Biometry Branch, National Cancer Institute, to Dr. M. Adrian Gross, Bureau of Drugs, FDA, February 22, 1972," all printed in *Hearing Before the Subcommittee on Health . . . July 20, 1972,* 100, 102, 103, 106, 125. Also see "DES Question: New Controversy," *Washington Post,* March 23, 1972, pp. B-1 and B-5, and "FDA Scientists Propose Ban on Cattle Growth Stimulant," *Washington Post,* March 23, 1972, p. G-7. The "virtual safety" concept was initially articulated in Nathan Mantel and W. Ray Bryan, "Safety Testing of Carcinogenic Agents," *Journal of the National Cancer Institute,* 27 (1961): 455–70.

17. "Leo Friedman, Ph.D., Director, Division of Toxicology, FDA, to Dr. Virgil O. Wodicka, Director, Bureau of Foods, FDA, December 17, 1971 and February 8, 1972," "Richard P. Lehmann, Ph.D., Director, Division of Nutritional Sciences, FDA, to Dr. C. D. Van Houweling, Director, Veterinary Medical Branch, FDA, December 21, 1971," and "J. Richard Crout, M. D., Deputy Director, Bureau of Drugs, FDA, to Henry E. Simmons, M. D., M. P. H., Director, Bureau of Drugs,

FDA, December 21, 1971," all printed in *Hearing Before the Subcommittee on Health . . . July 20, 1972,* 109, 117, 123, 133.

18. "Increase of DES Found," *Washington Post,* May 16, 1972, p. B-7; "Cancer Agent Rises in Meat," *Washington Post,* July 4, 1972, p. A-3; 72 U.S. Statutes 1784; "Cancer Institute Head Urges FDA Ban DES," *Washington Post,* July 10, 1972, p. A-2; and *Federal Register* 37 (June 21, 1972): 12251–53.

19. Public Law 87–781 and Public Law 90–399, 82 U.S. Statutes 342; William W. Goodrich, "The Food and Drug Administrations View on Procedural Rules," and William R. Pendergast, "Have the FDA Hearing Regulations Failed Us?" *Food Drug Cosmetic Law Journal* 23 (1968): 481–86, 524–31; and "New Drugs Regulations Under Drug Amendments of 1962," *Food Drug Cosmetic Law Reporter* (1963–67): 80107–09.

20. *Federal Register* 37 (August 4, 1972): 15747–50 and 37 (December 9, 1972): 26307; "U.S. Bans Production of Cattle Fattener," *Washington Post,* August 3, 1972, pp. A-1 and A-4; "Fattening Profits," *Nation* 215 (September 18, 1972): 196–97; Harrison Wellford, "Behind the Meat Counter," *Atlantic* 230 (October 1972): 86–90; "The Hormone Ban Spreads Confusion," *Business Week* (August 12, 1972): 29; and "How Much Is Too Much?" *New Republic* 167 (August 19 and 26, 1972): 7.

21. *Congressional Record,* 92nd Congress, 1st Session, 36259, 39734–35, and 40092, and 92nd Congress, 2nd Session, 26449–51, 29447, 29452, 29700, 30595–97, 31537–50, and 31853; "Report by the Committee on Government Operations Concerning Regulation of Diethylstilbestrol," *Food Drug Cosmetic Law Reporter* (1973–74): 40375–82; and Hess and Clark, Division of Rhodia, Inc. vs. Food and Drug Administration, et al., Vineland Laboratories, Inc. vs. Caspar W. Weinberger, Secretary, Department of Health, Education, and Welfare, et al., Chemetron Corporation and Dawes Laboratories, Inc. vs. the Department of Health, Education and Welfare, et al., and Hess and Clark, Division of Rhodia, Inc. vs. the Department of Health, Education and Welfare, et al., found in *United States Court of Appeals for the District of Columbia Circuit,* Docket Numbers 73–1581, 73–1589, 72–1864, and 72–2217.

22. *Food Drug Cosmetic Law Reporter* (1973–74): 40436–55.

23. "FDA Takes First Step to Comply with DES Order," *FDA Consumer* 8 (May 1974): 34; *Federal Register* 39 (March 27, 1974): 11299–301 and 39 (April 25, 1974): 14611; P. W. Aschbacher and E. J. Thacker, "Metabolic Fate of Oral Diethylstilbestrol in Steers," *J. Ani. Sci.* 39 (1974): 1185–92; and T. S. Rumsey, R. R. Oltjen, F. L. Daniels, and A. S. Kozak, "Depletion Patterns of Radioactivity and Tissue Residues in Beef Cattle After the Withdrawal of Oral 14C-Diethylstilbestrol," *J. Ani. Sci.* 40 (1975): 539–49. These studies may have been erroneous. See D. M. Tennent, R. F. Kouba, W. H. Ray, W. J. A. VanderHeuvel, and F. J. Wolf, "Impurities in Labeled Diethylstilbestrol: Identification of Pseudodiethylstilbestrol," *Science* 194 (1976): 1059–60.

24. *Federal Register* 39 (March 27, 1974): 11323–24; *Congressional Record,* 94th Congress, 1st Session, 5319–24, 27851–82, 28155–80, and 28337; "U.S. Cattle Meat Banned by Canada," *Washington Post,* April 10, 1974, p. C-6; "Answers on DES Program," *Cattleman* 62 (June 1975): 92, 94; "DES: Yes or No," *Feedlot Management* 16 (August 1974): 6–7; "Rules for Reporting DES-Free Livestock to Canada," *Journal of the American Veterinary Medical Association* 165 (1974): 603; and

"Canada Removes Ban on U.S. Livestock After Accord on DES," *Wall Street Journal,* August 5, 1974, p. 18.

25. *Federal Register* 41 (January 12, 1976): 1804–07 and (January 20, 1976): 2842; and John A. McLachlan, Retha R. Newbold, and Bill Bullock, "Reproductive Tract Lesions in Male Mice Exposed Prenatally to Diethylstilbestrol," *Science* 190 (1975): 991–92. Also see J. Carl Barrett, Annette Wong, and John A. McLachlan, "Diethylstilbestrol Induces Neoplastic Transformation Without Measurable Gene Mutation at Two Loci," *Science* 212 (June 19, 1981): 1402–04.

26. "Proposal to Withdraw Approval of the New Animal Drug Applications for Diethylstilbestrol," *Food Drug Cosmetic Law Reporter* (1978): 38784. For a brief history of CAST, see "The Voice of Agriculture Speaks Out," *Bioscience* 26 (1976): 481–84; B. P. Cardon, "Council For Agricultural Science and Technology," *Proceedings of the Agricultural Research Institute* 25 (1976): 23–29; Eliot Marshall, "Scientists Quit Antibiotics Panel at CAST," *Science* 203 (February 23, 1979): 732–33; and Gregg Hillyer, "No CASTing Lots to Inform Congress," *Iowa Agriculturist* 80 (Fall 1978): 14, 15, 19.

27. *Federal Register* 41 (November 26, 1976): 52105–07. The relevant CAST reports were "Efficiency in Animal Feeding with Particular Reference to Nonnutritive Feed Additives," CAST Report no. 22, January 18, 1974; and "Hormonally Active Substances in Foods: A Safety Evaluation," CAST Report no. 66, March 1977. For a retrospective of the FDA's assault on CAST, see Robin Marantz Hening, "CAST-Industry Tie Raises Credibility Concerns," *Bioscience* 29 (1979): 9–12 and 59. The hearings were held on January 5, February 14, May 16–20, May 23–27, July 5, and October 26–31, 1977.

28. For examples of the FDA's defense of its scientists' credibility, see, for instance, "FDA Director Suggests Formation of Advisory Panel on Drug Rules," *Washington Post,* December 14, 1972, p. A-2; "Industry Favoritism Allegations Convulse FDA," *Science and Government Report* 4 (September 15, 1974): 4; H. P. Eiduson and J. R. Weatherwax, "FDA Science Advisors," *FDA Papers* 1 (June 1967): 21–25; "Committee on Veterinary Medicine to Advice FDA," *Journal of the American Veterinary Medical Association* 148 (1966): 730–31; Edward G. Feldmann, "Diluting Science With Politics," *Journal of Pharmaceutical Sciences* 52 (January 1963): i; David H. Hickman, "Advisory Committee at FDA Legal Perspective," *Food Drug Cosmetic Law Journal* 29 (1974): 395–408; and "Shake-Up Asked for Drug Agency," *New York Times,* October 26, 1962, p. 8.

29. George E. Moore and William N. Palmer, "Money Causes Cancer: Ban It," *Journal of the American Medical Association* 238 (August 1, 1977): 397.

30. Jeremy J. Stone, "A Political Misuse of Science," *Bioscience* 28 (1978): 83; and "Filthy Lucre," *Wall Street Journal,* October 26, 1977, p. 22.

31. "FDA Seek to End DES Use for Animal Growth," *Chemical Week* 123 (October 4, 1978): 16; "It's Now Illegal to Use DES in Feedlot," *Wallaces Farmer* 104 (July 28, 1979): 34; "Proposal to Withdraw Approval of the New Animal Drug Applications for Diethylstilbestrol," *Food Drug Cosmetic Law Reporter* (1978): 38784–808; and *Federal Register* 44 (July 6, 1979): 39387–89 and 39618–19, and 44 (September 21, 1979): 54852–900. Congress's Office of Technology Assessment estimated that the banning of DES would cut beef supplies by about 4 percent and raise prices more than 9 percent. See "FDA Bans Most Uses of Controversial Drug," *Washington Post,* June 30, 1979, p. A-3.

32. "DES," *Cattleman* 66 (August 1979): 144; "Proposed DES Ban Postponed," *Des Moines Register,* August 2, 1979, p. A-7; "Rhone-Poulence, Inc., Hess and Clark Division V. FDA; Vineland Laboratories, Inc. V. FDA," *Food Drug Cosmetic Law Reporter* (1980–81): 38381–85; *Federal Register* 44 (July 20, 1979): 42679–80, and 44 (August 3, 1979): 45618 and 45764–65; "Iowa Cattle Didn't Get DES, Officials Say," *Des Moines Register,* April 5, 1980, p. A-3; "Iowa Veterinarians Probed for Possible Use of DES," *Des Moines Register,* April 29, 1980, p. A-3; and "The FDA Ban That Didn't Work," *Des Moines Register,* May 11, 1980, pp. F-1 and F-2.

33. Of use on the later twentieth century, see, for example, Alan I Marcus and Howard P. Segal, *Technology in America: A Brief History* (San Diego: Harcourt Brace Jovanovich, 1989), 315–60; William O'Neill, *Coming Apart: An Informal History of the 1960s* (Chicago: Quadrangle Books, 1971); Morris L. Dickstein, *Gates of Eden: American Culture in the 1960s* (New York: Oxford University Press, 1987); Shachtman, *Decade of Shocks: Dallas to Watergate, 1963–1974* (New York: Poseidon Press, 1983); and Christopher Lasch, *The Culture of Narcissism: American Life in an Age of Diminishing Expectations* (New York: Norton, 1978).

Contributors

Hamilton Cravens is professor of history, Iowa State University. Among his books are *The Triumph of Evolution: The Nature-Nurture Controversy, 1900–1941* (Baltimore: Johns Hopkins University Press, 1988 paper edition) and *Before Head Start: The Iowa Station and America's Children* (Chapel Hill: University of North Carolina Press, 1993). He has just finished a book on American culture in the early twentieth century and is now working on a book on the history of the social and behavioral sciences in Europe and America since the Enlightenment.

Robert B. Fairbanks is associate professor of history at the University of Texas at Arlington. Among his books are *Making Better Citizens: Housing Reform and the Community Development Strategy in Cincinnati, 1890–1960* (Urbana: University of Illinois Press, 1988) and *Essays on Sunbelt Cities and Recent Urban America* (College Station: Texas A&M University Press, 1990). He is also completing a book-length manuscript entitled "For the City as a Whole: Politics, Planning and the Public Interest in Dallas, 1920–1965."

David M. Katzman is professor of American Studies and of history at the University of Kansas. He is author of *Before the Ghetto: Black Detroit in the Nineteenth Century* (Urbana: University of Illinois Press, 1973), *Seven Days a Week: Women and Domestic Service in Industrializing America* (New York: Oxford University Press, 1978) and coauthor of *Plain Folk: The Life Histories of Undistinguished Americans* (Urbana: University of Illinois Press, 1982), *Three Generations in Twentieth Century America* (Homewood, Ill.: Dorsey, 1977), and

254   *Contributors*

A People and a Nation (Boston: Houghton Mifflin, 1990). Since 1977 he has been co-editor of *American Studies*.

Edwin T. Layton is professor of the history of science and technology at the University of Minnesota. He is the author of *Revolt of the Engineers: Social Responsibility and the American Engineering Profession* (Cleveland: Case-Western Reserve University Press, 1971), which won the Dexter Prize of the Society for the History of Technology, and numerous other books and papers, including *Technology and Social Change in America* (New York: Harper and Row, 1973). He has been president of the Society for the History of Technology and received the Leonardo da Vinci medal from that same society for his lifetime achievements. He is currently chair of the awards committee of the same society. His research interests include the interaction of science and technology, the history of water power, and engineering ethics.

M. Susan Lindee is assistant professor of the history and sociology of science at the University of Pennsylvania. She is the author of *Suffering Made Real: American Science and the Survivors at Hiroshima* (Chicago: University of Chicago Press, 1994) and (with Dorothy Nelkin) *Supergene: DNA in American Popular Culture* (San Francisco: W. H. Freeman, 1995).

Alan I Marcus is professor of history and director of the Center for Historical Studies at Iowa State University. He is author of *Agricultural Science and the Quest for Legitimacy: Farmers, Agricultural Colleges and Experiment Stations, 1870–1890* (Ames: Iowa State University Press, 1985); *Technology in America: A Brief History* (San Diego: Harcourt Brace Jovanovich, 1989); *Plague of Strangers: Social Groups and the Origins of City Services in Cincinnati* (Columbus: The Ohio State University Press, 1991); and *Cancer from Beef: DES, Federal Food Regulation, and Consumer Confidence* (Baltimore: Johns Hopkins University Press, 1994). His latest book is *Building Western Civilization: From the Advent of Writing to the Age of Steam* (San Diego: Harcourt, Brace, and Co., forthcoming, 1996).

Zane L. Miller is professor of American history and director of the Center for Neighborhood and Community Studies at the University of Cincinnati. He is the author of many books, including *Boss Cox's Cincinnati: Urban Politics in the Progressive Era* (New York: Oxford University Press, 1968), *The Urbanization of Modern America* (New York: Harcourt, Brace, and Co., 1973); *Suburb: Neighborhood and Community in Forest Park, Ohio, 1935–1976* (Knoxville: University of Tennessee Press, 1981); and *American Urbanism: A Historiographical Review* (Westport, Conn.: Greenwood, 1987). He is currently writing a history of the Cincinnati neighborhood of Clifton and a study of city planning and the city's Over-the-Rhine district since 1925. In addition, he is editor of the Urban Life and Urban Landscape Series of the Ohio State University Press and of the Greater Cincinnati Bicentennial History Series of the University of Illinois Press. He is also past president of the Urban History Association.

Judith Spraul-Schmidt is adjunct assistant professor in history, and lecturer in the evening college at the University of Cincinnati. She is on the City of Cincinnati Historic Conservation Board and the author of "Designing the Late Nineteenth Century Suburban Landscape: The Cincinnati Zoological Garden," and "Cultural Boosterism: The Construction of Cincinnati's Music Hall," which both appeared in *Queen City Heritage.*

Nancy M. Theriot is associate professor of history at the University of Louisville, where she is also chair of the Women's Studies program. Among her publications are *The Biosocial Construction of Femininity: Mothers and Daughters in Nineteenth Century America* (Westport, Conn.: Greenwood, 1988) and "Women's Voices in Nineteenth Century Medical Discourse: A Step Toward Deconstructing Science," *Signs* (1993).

Index

257